Teleportation

The Impossible Leap

DAVID DARLING

John Wiley & Sons, Inc.

Published by John Wiley & Sons, Inc., Hoboken, New Jersey
Published simultaneously in Canada

Design and composition by Navta Associates, Inc.

Library of Congress Cataloging-in-Publication Data:

Darling, David J.
Teleportation : the impossible leap / David Darling.
p. cm.
Includes bibliographical references and index.
ISBN-13 978-0-471-47095-3 (cloth)
ISBN-10 0-471-47095-3 (cloth)
1. Teleportation. 2. Quantum theory. I. Title.
BF1386.D37 2005
133.8–dc22

2004016045

Printed in the United States of America

10 9 8 7 6 5 4 3 2 1

To my granddaughter,
Emily Varie,
whose generation may see such wonders

Crazy way to travel–spreading a man's molecules all over the universe.

DR. McCOY, IN THE EPISODE TITLED "OBSESSION" OF THE *STAR TREK* ORIGINAL SERIES

Contents

Acknowledgments

For stimulating discussions and ideas, and also technical help, I'd like to thank Ignacio Cirac, Andrew Crayston, Nicolas Gisin, Fr. Gregory Hallam, Raymond Laflamme, David McMahon, Robert Nixon, the Oxford Centre for Quantum Computation, Jian-Wei Pan, Paramount Studios, Thomas Spitzer, Robert Weinberg, and Anton Zeilinger. Once again, I'm deeply grateful to senior editor Stephen Power and senior production editor Lisa Burstiner at John Wiley & Sons, and to my agent, Patricia Van der Leun. Last and most, thank-you, Jill, Lori-An, Jeff, and Mum and Dad for your loving support.

Prologue

Luk Barr entered the kiosk on Seventh Street, tapped a number into the keypad, and disappeared. A moment later, the gray damp of a New York evening gave way to bright morning sunlight streaming through the windows of his Sydney apartment, half a world away. Would he ever get used to this way of commuting?

He was ten when they teleported the first mammal–a white mouse–from one side of a lab to the other. Now, barely a quarter of a century later, trans-pads were as common as phone booths and shower stalls. Every home, office block, and public building had one; they were at the South Polar Base and on the moon–no one did the lunar run by spacecraft anymore. No rockets took off from Earth. Why bother when astronauts, supplies, and anything else could be teleported directly aboard a space station, and launched to other places from there? Ambulances were fitted with portable trans-pads so that the critically ill or injured could be beamed directly to whichever accident and emergency ward, anywhere in the world, had an open slot at the time. Computers automatically routed the patient and notified next of kin, who could then beam themselves to the hospital within seconds.

The world of the late twenty-third century had been revolutionized by the trans-pad. Of course, people still traveled by the old means, too, for pleasure or economy. Teleportation wasn't exactly cheap yet because of the vast amount of data involved. A typical jump would cost you the price of a fancy restaurant meal. But teleport service providers (TSPs) were cutting their rates all the time, as global data highways became ever broader. And that's what it was all about, this near-magical trick of teleportation. That's what you were turned into when you stood on the trans-pad: data.

Luk had never really understood the finer details of how teleportation worked, any more than most people had previously understood the innards of a TV or a CD player. It had to do with quantum properties of light and matter, and especially had something to do with a weird "spooky action at a distance" called entanglement. He also knew that teleporting complicated things like human beings hinged upon quantum computers.

The key was information. When you were teleported it wasn't in the form of matter or even energy. On the trans-pad, all the information was effectively sucked out of you and that process destroyed your material body, left you formless. Every atom of your substance was scanned, broken down, and sorted into its component elements. Carbon atoms went into one tank, oxygen into another, and so on. But you didn't get these atoms back. What was sent to the other end, at the speed of light, was just your pattern, your subatomic blueprint–pure information. At the destination, an identical copy of you was made from the atoms in the holding tanks there. And that always got Luk thinking. A moment ago he'd consisted of *his* atoms. Now here he was, someplace else, built up from the stuff of someone else's old body. But what the heck, one carbon atom was like any other.

Not everyone agreed with that. Luk knew one or two "originals"—people who'd never jumped and never would. Some had a morbid fear of it, like the fear of flying. Others were afraid for a different, more disturbing reason: they were sure they wouldn't survive the trip. That's because they believed *no one* survived the trip. Teleportation was the same as dying. The identical copy at the other end was exactly that—a copy, a clone. But it wasn't you. That person had ceased to exist.

Luk had once gotten into conversation with a fundamentalist original.

"Your soul doesn't make the jump you know," this guy had said.

"I'm not sure I believe in souls," Luk had replied. "In any case, I'd have thought that if you're copied atom for atom, whatever other stuff you're made of gets copied over too."

"That's where you're wrong. But it's too late to do anything about it now. Not even worth praying about, because without a soul . . ."

It hadn't really bothered Luk at the time. He'd heard all about these fundamentalists and their churches from which "jumpers"—the vast majority of the population—were excluded. But then he started thinking more about it. What if there was something about you that didn't make the transfer? On the other hand, without jumping there'd be so much he couldn't have done. Like being with his family and friends, instantly, anyplace, anytime. Like visiting any point on Earth at the drop of a hat. Only yesterday, he and Rosie had spent a couple of hours at the temple of Angkor Wat in Cambodia; last week, it had been an afternoon's trek in the Himalayan foothills and back home for supper. No, he couldn't imagine life the old way anymore. But he really ought to try to get his head around the science of the thing.

Rosie was already up and about, pouring some fresh coffee. Three great sights in one, if you counted in the Harbor Bridge through the glass wall beyond.

"Say, Rosie. Any idea where that book on teleportation got to?"

Introduction

A Brief History of Beaming Up

The idea of people and things vanishing from one place and mysteriously reappearing in another, possibly having passed through solid walls, goes back thousands of years. Stories of ghosts and spirits are at least as old as civilization, and probably very much older. Reports of objects and animals suddenly popping up where they're least expected also have a long history.

In Oriental mysticism and Western occultism there's the notion of *apport*: an object or person winking out in one place and, driven by some unknown mental power, materializing somewhere else, perhaps far away. The Buddha reputedly vanished from India and reappeared shortly after in Sri

Lanka. Devotees of the Indian yogi Satya Sai Baba hail him as the current king of apport. Other instances of supernatural transport crop up in the Bible, including one in Acts 8:39–40: "The Spirit of the Lord caught away Philip. . . . But Philip was found at Azotus." Also high on the list of odd comings and goings is *bilocation*–the phenomenon of being in two places at once–which is talked about in Catholic philosophy. Several Christian saints and monks were supposedly adept at this, including St. Anthony of Padua, St. Ambrose of Milan, and Padre Pio of Italy. It's said that in 1774 St. Alphonsus Maria de'Ligouri was seen at the bedside of the dying Pope Clement XIV, when in fact the saint was confined to his cell four days' journey away.

Many such tales of strange materializations were collected by that famed purveyor of the weird and unorthodox, Charles Hoy Fort, a native of Albany, New York. Fort was something of a curiosity himself: a small, round, Wellesian figure with a luxuriant walrus moustache, whose favorite haunt was the New York Public Library. His pockets stuffed with index cards, pencils, and other flotsam of his private research, he'd trundle through the aisles scouring books and journals for any murmurings of the unconventional, the iconoclastic. Then back he'd ride, on the uptown trolley, to his wife and home in the Bronx, to his modest apartment with its crowded rooms, its cozy kitchen, its little office bulging with files of wonders such as no one had ever assembled before. Fish and frogs falling out of clear blue skies, stones flowing in open fields, and other troublesome oddities filled his vast catalog of the anomalous. These "Forteans," as they've become known, were compiled by the little investigator into several books, beginning with *The Book of the Damned* in 1919. In his third published collection of the strange and mysterious, *Lo!* (1931), Fort coined a new term: "Mostly in this book I shall specialize upon indications that there exists a transportory force that I shall call *Teleportation.*"

To Fort, teleportation was the master link that underpinned his arcane world of incongruities. It was nature's trickster force that could conjure up a poltergeist or make a holy statue bleed, that could shower living things from a cloudless sky, and that lay behind what was then the unfathomable breeding habits of eels. Nothing was solid in Fort's view: our present surroundings are a mere quasi-existence, a twilight zone between many different layers of reality (whatever that might be) and unreality. Teleportation was the means by which the contents of another level of existence, normally hidden from view, could suddenly cross over and intrude, seemingly from nowhere, into our own plane.

Fort explored the shadowlands between established fact and myth, between the known and the fantastic. But during the golden age of science fiction, from the late 1930s to the fifties, and in a few cases much earlier than that, more hard-edged, technological ideas about teleportation began to emerge. Authors pondered the physics by which a person might hop around the Galaxy in disembodied form. If nothing else, being able to whisk characters effortlessly between worlds in the wink of an eye was a handy way of streamlining plots that might otherwise get mired in lengthy interplanetary or interstellar journeys.

Perhaps the earliest recorded story of a matter transmitter was Edward Page Mitchell's "The Man Without a Body" in 1877. In this story, a scientist invents a machine that breaks down the atoms of a cat and sends them by wire to a receiver, where the animal is reassembled alive and well. The inventor then tries it upon himself, but unfortunately the battery runs down before he can transmit more than just his head.

Arthur Conan Doyle was also early on the teleportation scene with one of his Professor Challenger yarns, "The Disintegration Machine," first published in 1927, several years after Fort coined the term (which doesn't actually appear in the story). The evil Nemor, who has built the device, asks Challenger, "Can you conceive a process by which you, an organic being, are in some way dissolved into the cosmos, and then by a subtle reversal of the conditions reassembled once more?" Nemor explains that all things have an invisible framework that draws the atoms back into place. "How can such a thing be done," says Nemor, "save by the loosening of the molecules, their conveyance upon an etheric wave, and their reassembling, upon exactly its own place, drawn together by some irresistible law?"

In *Special Delivery* (1945), George O. Smith describes how transmitters scan an object atom by atom and then take it apart, storing the particles in a "matter bank." The information and energy released during the breakup are then beamed to a second station that uses raw materials in its own matter bank to recreate the body perfectly, down to the last atom. In *The Mixed Men* (1952), E. A. Van Vogt describes two ways to teleport: people can be sent through space as electronic images–elaborate, three-dimensional faxes–to be reassembled at the receiving end from local organic material, or they can be turned into a flow of electrons and then converted back into their normal atomic form on arrival. Alfred Bester's *The Stars My Destination* (1956) posits a whole culture based on a form of teleportation called "jaunting" that, while never properly explained, involves the jaunter clearly envisioning the place where he or she hopes to materialize. Algis

Budrys's haunting 1960 novel *Rogue Moon* saw travelers making the hop from Earth to the moon in data form and being reconstructed from fresh matter at the lunar end. Even cartoon superheroes got in on the act. Several of the mutant X-Men from Marvel Comics were able to teleport in different ways. Nightcrawler, for instance, would "bamf" over short distances, leaving the scent of brimstone in his wake and taking along anyone who happened to be touching him at the time.

Not only did teleportation afford writers a quick way of getting their characters from A to B, but the perils and possibilities of the device itself offered new narrative potential. George Langelaan's short story "The Fly" explored the dire consequences of a teleportation gone wrong. It first appeared in the June 1957 issue of *Playboy* magazine before being turned into a screenplay by James Clavell, which formed the basis of the classic 1958 horror film *The Fly* directed by Kurt Neumann. Scientist Andre Delambre (played by David Hedison), while researching matter transmission, suffers the gruesome effects of dematerializing at the very moment a fly enters the teleportation chamber. His molecules mingle with those of the fly, and he begins a slow and agonizing transformation into a giant insect. Eventually unable to talk, he types notes to his equally beleaguered wife, Helene (Patricia Owens), as she desperately and unsuccessfully searches for the unique, human-headed fly needed to reverse the metamorphosis.

By the 1960s, "transfer booths" for personal international and interplanetary jet-setting had become sci-fi de rigueur. Clifford Simak put them to good use in his novel *Way Station* (1964), while Larry Niven featured them in his award-winning book *Ringworld* (1970). And then, of course, there was *Star Trek*.

The original three-season TV series, which ran from 1967 to 1969, brought, along with warp factors, Klingons, and green-skinned go-go girls, the idea of teleportation into everyone's living room. "Beam me up, Scotty" became one of the small screen's most oft-repeated lines (though trivia hunters will find that "Beam me up, Mr. Scott" is the closest the show actually came to that immortal line). In the twenty-third century world of Kirk, Spock, and McCoy, shuttlecraft are used only in special circumstances–ion storms are a favorite–when beaming someone's molecules around might prove a health hazard. But, ironically, the reason that *Trek* mastermind Gene Roddenberry chose to equip his starships with "transporters" had less to do with high-tech future possibilities than with low-tech Beatles-era reality. It wasn't feasible, in terms of budget or sixties-level special effects, to show convincingly a spacecraft landing on a different planet every week. Much easier to have crewmembers shimmer out in one scene, then twinkle back an instant later someplace else. With realistic computer graphics still a couple of decades away, the effect called for plenty of ingenuity and homespun improvisation. The sparkling dematerialization and rematerialization sequences were created by dropping tiny bits of aluminum foil and aluminum perchlorate powder against a black sheet of cardboard, and photographing them illuminated from the side by a bright light. When the characters were filmed walking into the transporter, they stepped onto the pads, Kirk gave the order to energize, and the actors stepped off. In the studio lab, after the film was developed, the actors were superimposed fading out and the fluttering aluminum fading in, or vice versa. By 1994, when production started on the fourth TV incarnation of the franchise, *Star Trek: Voyager*, computer graphics was well into its stride and a new transporter effect was devised in which little spheres of light expand to cover the person, a shower of falling glitter providing a nod to the past.

The universe of *Star Trek* may be only make-believe. The staff at Paramount may have no more idea how to beam a person around than Leonard Nimoy has of performing an actual mind-meld. But the *Trek* transporter has brought the notion of teleportation into millions of homes worldwide, and given us a common set of images and expectations. Over the course of hundreds of episodes, the transporter's technical specs have been fleshed out and its dramatic possibilities explored in more detail than almost any other device in the history of science fiction.

According to the official bible of Trekana, *The Star Trek Encyclopedia*, the transporter "briefly converts an object or person into energy, beams that energy to another location, then reassembles the subject into its original form." A little short on detail perhaps for those interested in cobbling together a version of their own to avoid the daily rush hour, but no matter: when facts are hard to come by, there's always technobabble to fill the void.

A key element of the *Trek*-style transporter is the so-called annular confinement beam (ACB), a cylindrical force field that channels and keeps track of the transportee from source to destination. Basically, this stops your bits and pieces from drifting off into interstellar space while you're being dispatched to the surface of some strange new world. It seems that the ACB first locks onto and then disassembles the subject into an energy- or plasmalike state, known as phased matter. This is a key step in the whole process, so it's unfortunate that the show's creators can't be a little more specific (and win a Nobel Prize while they're at it). But what's clear is that some "stuff," be it matter or energy or some hybrid of these, is sent from one place to another, along with instructions needed to reconstitute the subject upon arrival. George O. Smith would have been delighted that his *Special Delivery* system, or something very much like it, eventually found its way into Hollywood's most celebrated starship.

Imagine, then, that you've stepped onto the transporter pad, issued the fateful command "energize," and had your atoms turned into phased matter. Now you're all set to go. Your matter stream is fed into a pattern buffer (a hyperlarge computer memory that briefly stores your entire atomic blueprint), piped to one of the beam emitters on the hull of the starship, and then relayed to a point on the ground where, if your luck holds, the ACB will put you back together again. There's even a component of the transporter designed to shimmy around one of the most basic laws of physics, known as Heisenberg's uncertainty principle. This frustrating little rule insists that you can never know exactly where something is and exactly how it's moving at the same time. Unnoticeable in the everyday world, it comes into effect with a vengeance at the subatomic level and, at first sight, seems to pose one of the biggest obstacles to practical teleportation. How can an exact copy of you be made somewhere else if it's impossible to establish the state of every particle in your body at the outset? No problem, according to Mike Okuda, the scenic art supervisor for the *Star Trek* spinoffs *Deep Space Nine*, *Voyager*, and *Enterprise*. His answer to Heisenberg's uncertainty principle: the Heisenberg compensator. (Once asked how it worked, Okuda replied, "Very well, thank you!")

Anyone wondering whether he or she would have the guts to step up to the transporter plate along with the other crewmembers and be boldly sent needs to bear two thoughts in mind. First, for reasons we'll go into later in the book, teleportation could probably never work along the lines just described (hint: a "Heisenberg compensator" is physically impossible). Second, even in the *Star Trek* universe, transporters can go wrong. Well, *of course* they can go wrong—that's part of the fun.

One (or two) of William Shatner's better performances as Kirk came in *Star Trek*'s first-season episode, "The Enemy Within," written by the top-drawer science fiction author Richard Matheson, who also penned some of the more memorable episodes of *The Twilight Zone* (including "Nightmare at 20,000 Feet" in which Shatner sees a gremlin on the wing of a plane). Having beamed up from a mission on a faraway planet, Kirk feels faint and is helped from the transporter room by Mr. Scott. A moment later a duplicate Kirk appears on the pad. Apparently the magnetic effects of an ore on the planet's surface interfered with the transporter and caused it to split the captain into two selves: one good but incapable of making decisions, the other evil and strong-willed. In this interesting twist on the Jekyll and Hyde theme, it becomes clear that the two halves can't survive apart and that the violent, animal-like component is just as essential in making Kirk an effective leader as his benign side.

Transporter fission turns to fusion in the *Voyager* episode "Tuvix," when crewmates Tuvok, the Vulcan security officer, and Neelix, the spotty Talaxian, longtime antagonists, are merged during a teleportation into one person. The resulting Tuvix harbors the memories of both progenitors but has a single consciousness. Initially confused and ambivalent, Tuvix eventually carves out a clear identity and personality of his own, and when a means is discovered to undo the mix-up caused by the transporter accident, he objects, not unreasonably, on the grounds that it will kill him. Captain Janeway is faced with the moral dilemma of either ending the brief existence of a distinct, unique individual who has become well-liked among the crew, or denying the rights of Tuvok and Neelix to resume their separate lives. Ensemble casting and contractual arrangements being what they are, Tuvix is consigned to oblivion.

All good fun, of course–and useful grist for the philosophical mill. But in 1993, as *Star Trek* began its third incarnation, *Deep Space Nine*, something happened in the real universe to make beaming up seem just a little less fantastic: plans were published for building the first practical teleporter.

Today, far from being a science fiction dream, teleportation happens routinely in laboratories all around the world. It isn't as dramatic as its *Star Trek* counterpart–yet. No one has had his or her atoms pulled apart in Seattle and been reconstituted moments later in Seville. The researchers doing this sort of thing aren't mad scientists intent on beaming the molecules of unfortunate animals around the lab and hoping for the best. Instead, real teleporteers belong to a group of computer specialists and physicists who share a common interest. All are involved, in one form or another, with tackling the same questions: How can information be handled at the smallest level of nature? How can messages and data be sent using individual subatomic particles?

Teleportation in the real world means *quantum* teleportation. Working at the quantum level, it turns out, is the only way to make an exactly perfect copy of the original. So, to understand how teleportation works means taking a trip into the weird world of quantum mechanics. It means looking at how light and matter behave at an ultra-small scale, where extraordinary things are commonplace, and common sense goes out the window.

Actual teleportation, as it's done at present, doesn't involve a flow of matter or energy. It doesn't work by streaming atoms, or any other physical "stuff," from one place to another like the Enterprise's transporter. The basis of true teleportation is transferring *information* without sending it through ordinary space. It's a transfer achieved with the help of the strangest, most mysterious phenomenon in all of science: entanglement. A bizarre shifting of physical characteristics between nature's tiniest particles, no matter how far apart they are, entanglement lies at the heart of teleportation as well as two other major new fields of research: quantum cryptography and quantum computation. As we'll discover, while teleportation is an active area of research in and of itself, it's closely related to the art of sending secret codes and is vitally important in building quantum computers–potentially the most powerful information processors the world has ever seen.

For now, the most we can teleport is light beams, subatomic particles, and quantum properties of atoms, rather than solid objects. But scientists are talking about teleporting molecules sometime within the next decade. Beyond that there's the prospect of doing the same with larger inanimate things. And beyond *that* . . .

1

Light Readings

It was a low-key affair: no television cameras or reporters were on hand. No audience of millions was hanging breathlessly on the mission's success. In a small, darkened room at the University of Innsbruck in 1997, on a lab bench strewn with cables and electro-optical gear, scientists destroyed a few bits of light in one place and made perfect replicas about a meter away. True, it lacked the drama and visceral appeal of Armstrong and Aldrin's exploits on the moon. But in the long run it was no less important. The first teleportation in history had just taken place.

Nothing is more familiar than light. It's the moon reflected on a calm ocean, a sparkling diamond, a rainbow, a glowing ember. We take it for granted because it's all around us. But behind everyday appearances lies the question of what light really is–what it's made of, how it behaves, and how on Earth it's possible to make it vanish in one place and then reappear, an instant later, somewhere else. To understand the science of teleportation we first have to come to grips with the nature of light.

In ancient Greek mythology, the goddess Aphrodite fashioned the human eye out of the four elements–earth, air, fire, and water–then lit a flame inside the eye to shine out and make sight possible. According to this explanation, we can't see in the dark because rays from the eyes must interact with rays from a source such as the sun. So taught Empedocles, in the fifth century B.C., and, for most Europeans, it was a theory good enough to stand for the next two thousand years.

Other ancients, though, came closer to the modern view. The Roman poet and philosopher Lucretius was far ahead of his time when, in 55 B.C., he wrote: "The light and heat of the sun; these are composed of minute atoms which, when they are pushed off, lose no time in shooting right across the interspace of air in the direction imparted by the push." Earlier, Euclid had described the laws of reflection and argued that light travels in straight lines. In about A.D. 140, Ptolemy found, from careful measurements of the positions of stars, that light is refracted, or bent, as it passes through the atmosphere.

Most advanced of all, before the scientists of the Renaissance, was Ibn al-Haytham, who lived around the turn of the tenth century in what is now Iraq. He rejected the eye-beam idea, understood that light must have a large but finite velocity, and realized that refraction is caused by the velocity being different in different substances. Those facts had to be relearned in the West several centuries later. But when they had been, a great debate sprang up around light's basic nature that set the stage for the startling revelations of more recent times.

By the seventeenth century, two theories competed to explain the underlying essence of light. Isaac Newton insisted that light is made of particles or, to use the term then in vogue, *corpuscles* (meaning "little bodies"). His contemporary, the Dutch physicist Christiaan Huygens, championed the idea that light consists of waves.

Newton's corpuscular theory fit in well with some of his other groundbreaking work on the way objects move. After all, light is seen to travel in straight lines, and how it is reflected by a mirror seems similar to how a ball bounces off a wall. Newton revolutionized optics: he split apart white light with prisms and showed that it's a mixture of all the colors of the rainbow, and he built the first reflecting telescope. But all his thoughts in this field were guided by his belief that light is a stream of little particles.

Huygens took the rival view that light is really made of waves, like those that ripple out when a stone is tossed into a lake. The medium through which light waves travel, Huygens supposed, is an invisible substance called the luminiferous ether–an idea he inherited from his old tutor René Descartes.

Both corpuscular and wave theories could explain perfectly well the reflection and refraction of light. It's true Newton and Huygens differed in their predictions about the way the speed of light changes as light goes from a less dense medium, such as air, to a more dense one, such as glass; Newton said it should go up, and Huygens believed it should go down. As there was no way of measuring this speed change at the time, however, it couldn't be used as an experimental test. One observation, though, did tilt the scales of seventeenth-century opinion. When light from a faraway source, such as the sun, passes a sharp edge, such as the wall of a house, it casts a sharp-edged shadow. That's exactly what you'd expect of streams of particles, traveling on dead straight courses. On the other hand, if light were made of waves it ought to diffract, or spread around corners, just as ocean waves wash around the sides of a harbor wall, and cast a shadow that was fuzzy-edged. The observation of clean-edged shadows, together with Newton's huge stature in science, guaranteed almost unanimous support for the corpuscular view. Then came a sea change. Early in the nineteenth century, the balance of opinion started to shift emphatically the other way, and the wave picture of light moved to center stage.

The instigator of this shift was Thomas Young, an English physician, physicist, and linguist extraordinaire. The first of ten children of Quaker parents, Young was a precocious youngster of fiercely independent mind who learned to read at the age of two, knew Latin as a six-year-old, and was fluent in thirteen languages while still a teenager. Later he played a key role in unraveling the mysteries of Egyptian hieroglyphics through his deciphering of several cartouches—oval figures containing royal names—on the Rosetta stone. But his greatest claim to fame lies with his work on optics.

Having studied medicine in London, Edinburgh, Göttingen, and Cambridge, Young bought a house in London with money left him by a wealthy uncle and set up practice there. From 1811 to the time of his death, he served as a physician at St. George's Hospital. His main medical interest, though, wasn't in treating patients but in doing research.

Human vision and the mechanism of the eye held a special fascination for him.

As early as 1790 or thereabouts, barely out of school, Young had hatched the original theory of how color vision comes about, building on work by Newton. Through his experiments with prisms back in 1672, Newton had shown that, rather surprisingly, ordinary white light is a thorough blend of all the rainbow colors from red to violet. Objects have a particular hue, Newton realized, because they reflect some colors more than others. A red apple is red because it reflects rays from the red end of the spectrum and absorbs rays from the blue end. A blueberry, on the other hand, reflects strongly at the blue end of the spectrum and absorbs the red. Thinking about Newton's discovery, Young concluded that the retina at the back of the eye couldn't possibly have a different receptor for each type of light, because there was a continuum of colors from red to violet. There was no way there could be such a vast number of specific receptors. Instead, he proposed that colors were perceived by way of a simple three-color code. As artists knew well, any color of the spectrum (except white) could be matched by judicious blending of just three colors of paint. Young suggested that this three-color code wasn't an intrinsic property of light, but arose from the combined activity of three different "particles" in the retina, each sensitive to different parts of the spectrum. In fact, we now know that color vision depends on the interaction of three types of cone cells: one especially sensitive to red light, another to green light, and a third to blue light. Considering that Young set out his three-color theory before cone cells had been discovered, he came remarkably close to the truth.

While still a medical student at Emmanuel College in

Cambridge, Young also discovered how the lens of the eye changes shape to focus on objects at different distances. In 1801, just after his move to London, he showed that astigmatism results from an abnormally shaped cornea. At the age of only twenty-eight, Young was already professor of natural philosophy at the Royal Institution and lecturing on just about everything under and above the sun: acoustics, optics, gravitation, astronomy, tides, electricity, energy (he was the first to give the word *energy* its scientific significance), climate, animal life, vegetation, cohesion and capillary attraction of liquids, and the hydrodynamics of reservoirs, canals, and harbors. His epitaph in Westminster Abbey says it all: "a man alike eminent in almost every department of human learning."

It was Young's work on optics that eventually made him famous–and a heretic in his own land. Having pioneered physiological optics it was only a short step to considering the fundamental essence of light, and in that fateful year of 1801, Young turned his mind to light's basic nature. His interest in this question was piqued by some work he'd done in the mid-1790s on the transmission of sound, which he came to believe was analogous to light. Sound was made of waves. Young suspected that light was, too. So in 1802, he devised an experiment to put this theory to the test.

If light were made of waves, then from a very narrow opening in its path it should head out as a series of concentric, circular ripples. To grasp this idea, imagine a long rectangular trough of water. Halfway down the trough is a barrier with a small hole in it. Straight, parallel waves, like the lines of waves marching toward a shore before they break, are created by moving a plank of wood back and forth at one end of the trough. When a wave reaches the barrier it is stopped dead in its tracks–except for at the small hole.

This opening serves as a new source of waves, but of expanding circular waves, as if a pebble had been dropped into the water at that point. On the side of the tank beyond the barrier, the secondary waves fan out, circles within circles. Now suppose there are *two* little holes in the barrier across the water tank. Both act as sources of circular waves. What's more, these waves are exactly in step–in *phase*, to use the scientific description–because they've come from the same set of waves that arrived at the barrier. As the circular waves spread out from the two holes, they run into one another and interact. They *interfere*. Where two crests or two troughs coincide, they combine to give a crest or trough of double the height. Where a crest meets a trough, the two cancel out to leave an undisturbed spot. Drop two pebbles of equal size close together in a pond and you'll see an instant demonstration. The result is an interference pattern.

Thomas Young carried out the equivalent of this water-wave interference experiment using light. In a darkened room, he shone light upon a barrier in which there were two narrow, parallel slits, within a fraction of an inch of each other. Then he looked at the outcome on a white screen set farther back. If light were made of particles, as Newton claimed, the only thing showing on the screen ought to be two bright parallel lines where the light particles had shot straight through. On the other hand, if light, like water, were wavelike, the secondary light waves spreading out from the two slits should create a pattern of alternate dark and light bands, where the light from the two sources respectively canceled out and amplified. Young's result was literally black and white: a series of interference bands. His double-slit experiment argued powerfully in favor of the wave model of light.

Flushed with success, Young used his proof of the wave character of light to explain the beautiful, shifting colors of thin films, such as those of soap bubbles. Relating color to wavelength, he also calculated the approximate wavelengths of the seven colors of the rainbow recognized by Newton. In 1817, he proposed that light waves were transverse, in other words that they vibrate at right angles to the direction in which they travel. Up to that time, supporters of the wave theory of light had assumed that, like sound waves, light waves were longitudinal, vibrating along their direction of motion. Using his novel idea of transverse waves, Young was able to explain polarization–a phenomenon that looms large in our story of teleportation–as the alignment of light waves so that they vibrate in the same plane.

For these breakthroughs, Young ought to have been hailed as a wunderkind, a youthful genius who set physics on its head. But he'd had the audacity to challenge the authority of Newton in the great man's own domain, and that was the scientific equivalent of hari-kari. A savage, anonymous review of Young's work in 1803 in the *Edinburgh Review* (now known to have been by Lord Henry Brougham, a big fan of the corpuscular theory) cast Young into scientific limbo for at least a decade. Newton had been dead eighty years when Young officially published his findings on interference in 1807.[51] But the godlike status of the great man in Britain meant that Young's compelling results were pretty much ignored by his compatriots.

Instead it fell to a Frenchman, Augustin Fresnel, to persuade the world a few years later, through a series of demonstrations that were more comprehensive than those of Young, that light really was a series of waves and not a movement of minuscule particles. By the mid-nineteenth century, when

Léon Foucault, another French physicist, showed that, contrary to the expectations of Newton's corpuscular theory, light traveled more slowly in water than in air, the wave picture of light was firmly established. All that remained was to clear up some details. If light consisted of waves, what was the nature of these waves? What exactly was waving?

In the 1860s, the final pieces of the puzzle of light seemed to fall into place thanks to the work of the Scottish theoretical physicist James Clerk Maxwell.* What Maxwell found was a set of relationships–four simple-looking formulas (though not so simple in fact!) known ever since as Maxwell's equations–that bind electricity and magnetism inextricably together. These equations explained, for example, all the results of the pivotal experiments on electric currents and magnetic fields that Michael Faraday had carried out in the 1830s at the Royal Institution in London. Maxwell's equations assume that around magnets, electric charges, and currents there exist regions of influence known as fields. Crucially, the equations show that it's meaningless to talk about these fields on their own. Wherever there's an electric field, there's always an accompanying magnetic field at right angles to it, and vice versa. The two can't exist apart, and together they produce a marriage called an electromagnetic field.

As Maxwell's equations make clear, change the electric field and the magnetic field responds by changing as well. The disturbance to the magnetic field then causes a further shift in the electric field, and so on, back and forth, action

---

*Maxwell's original name was James Clerk. "Maxwell" was added to his surname after his father inherited an estate near Edinburgh from his Maxwell ancestors.

and reaction. Maxwell was aware of the implication of this two-way feedback. It meant that any fluctuation in an electric or magnetic field gives rise to electromagnetic waves. The frequency of these waves equals the rate at which the electromagnetic field waxes and wanes. Using his formulas, Maxwell could even figure out the speed at which electromagnetic waves should travel: 300,000 kilometers per second. But this value was already very familiar in science. It was equal to the speed of light, which by the mid-nineteenth century was known quite accurately, and there was no way Maxwell was going to take that as mere coincidence. In 1867, he proposed that light waves were none other than electromagnetic waves. "We can scarcely avoid the conclusion," he wrote, "that light consists in the transverse undulations of the same medium which is the cause of electric and magnetic phenomena."

Nor did it end with visible light. Maxwell's equations implied that there ought to be a broad span of electromagnetic radiation, stretching from very long wavelengths to very short ones. (After all, there was no limit, in principle, to the rate at which an electric charge or a magnet could vibrate.) Think of the electromagnetic spectrum as being like the keyboard of a piano, with the rainbow of visible light, from red to violet, akin to a single octave somewhere in the middle. Sadly, Maxwell didn't live to see his prediction that there must be other types of electromagnetic radiation bear fruit, for he died of abdominal cancer in 1879 at only forty-eight. What's more, as in the case of Thomas Young, his discoveries were never properly appreciated during his lifetime, especially by his compatriots. (As did Young, Maxwell did important work on color vision, especially color blindness, and was a pioneer of color photography.) Most contemporary scientists, including Lord Kelvin (William Thomson), the top scientific dog of the day, didn't accept Maxwell's theory of electromagnetism, and not many understood the mathematics behind his equations. In Britain, he had the support of only a small circle of young scientists.

Nine years after Maxwell's death, however, the truth and power of his equations was borne out. The German physicist Heinrich Hertz found notes at the low (long wavelength) end of the electromagnetic keyboard in the form of radio waves, which he produced by making sparks fly back and forth between two brass electrodes. In 1895 another German, Wilhelm Röntgen, discovered what turned out to be radiation from the short wavelength end of the spectrum–X-rays.

By the close of the nineteenth century, the particle theory of light looked to be dead in the water. Maxwell's electromagnetic theory had been fully vindicated and there seemed no longer to be any doubt that light was purely a form of wave motion. But then something extraordinary happened: just when scientists were beginning to feel smug about having all the cosmic laws in their grasp, cracks began to appear in the vast fortress of Victorian physics–experimental results that defied explanation in terms of known science. Minor at first, these cracks widened until it became clear that the old scientific edifice, seemingly so secure, would have to be torn down to its foundations and replaced–by something new and disturbingly alien.

The greatest crisis physics has ever known came to a head over afternoon tea on Sunday, October 7, 1900, at the home of Max Planck in Berlin. The son of a professor of jurisprudence, Planck had held the chair in theoretical physics at the

University of Berlin since 1889, specializing in thermodynamics–the science of heat change and energy conservation. He could easily have been sidetracked into a different career. At the age of sixteen, having just entered the University of Munich, he was told by Philipp von Jolly, a professor there, that the task of physics was more or less complete. The main theories were in place, all the great discoveries had been made, and only a few minor details needed filling in here and there by generations to come. It was a view, disastrously wrong but widely held at the time, fueled by technological triumphs and the seemingly all-pervasive power of Newton's mechanics and Maxwell's electromagnetic theory. Planck later recalled why he persisted with physics: "The outside world is something independent from man, something absolute, and the quest for the laws which apply to this absolute appeared to me as the most sublime scientific pursuit in life." The first instance of an absolute in nature that struck Planck deeply, even as a high school student, was the law of conservation of energy–the first law of thermodynamics. This law states that you can't make or destroy energy, but only change it from one form to another; or, to put it another way, the total energy of a *closed* system (one that energy can't enter or leave) always stays the same.

Later, Planck became equally convinced (but mistakenly, it turned out) that the second law of thermodynamics is an absolute as well. The second law, which includes the statement that you can't turn heat into mechanical work with 100 percent efficiency, crops up today in science in all sorts of different guises. It forms an important bridge between physics and information–a bridge built in part, as we'll see, by one of the pioneers of teleportation. When the second law was first introduced in the 1860s by Rudolf Clausius in Germany and Lord Kelvin in Britain, however, it was in a form that came to be known as the entropy law. Physicists and engineers of this period were obsessed with steam engines, and for good reason. Steam engines literally powered the Industrial Revolution, so making them work better and more efficiently was of vital economic concern. An important early theoretical study of heat engines had been done in the 1820s by the French physicist Sadi Carnot, who showed that what drives a steam engine is the fall or flow of heat from higher to lower temperatures, like the fall of a stream of water that turns a mill wheel.

Clausius and Kelvin took this concept and generalized it. Their key insight was that the world is inherently active, and that whenever an energy distribution is out of kilter, a potential or thermodynamic "force" is set up that nature immediately moves to quell or minimize. All changes in the real world can then be seen as consequences of this tendency. In the case of a steam engine, pistons go up and down, a crank turns, one kind of work is turned into another, but this is always at the cost of a certain amount of waste heat. Some coherent work (the atoms of the piston all moving in the same order) turns into incoherent heat (hot atoms bouncing around at random). You can't throw the process into reverse, any more than you can make a broken glass jump off the floor and reassemble itself on a table again. You can't make an engine that will run forever; the reason the engine runs in the first place is because the process is fundamentally unbalanced. (Would-be inventors of perpetual motion machines take note.)

Whereas the first law of thermodynamics deals with things that stay the same, or in which there's no distinction between past, present, and future, the second law gives a motivation for change in the world and a reason why time seems to have a definite, preferred direction. Time moves relentlessly along the path toward cosmic dullness. Kelvin spoke in doom-mongering terms of the eventual "heat death" of the universe when, in the far future, there will be no energy potentials left and therefore no possibility of further, meaningful change.

Clausius coined the term *entropy* in 1865 to refer to the potential that's dissipated or lost whenever a natural process takes place. The second law, in its original form, states that the world acts spontaneously to minimize potentials, or, equivalently, to maximize entropy. Time's arrow points in the direction the second law operates–toward the inevitable rise of entropy and the loss of useful thermodynamic energy.

For Max Planck, the second law and the concept of entropy held an irresistible attraction–the prospect of an ultimate truth from which all other aspects of the external world could be understood. These ideas formed the subject of his doctoral dissertation at Munich and lay at the core of almost all his work until about 1905. It was a fascination that impelled him toward the discovery for which he became famous. Yet, ironically, this discovery, and the revolution it sparked, eventually called into question the very separation between humankind and the world, between subject and object, for which Planck held physics in such high esteem.

Planck wasn't a radical or a subversive in any way; he didn't swim instinctively against the tide of orthodoxy. On the contrary, having come from a long line of distinguished and very respectable clergymen, statesmen, and lawyers, he tended to be quite staid in his thinking. At the same time, he also had a kind of aristocratic attitude toward physics that led him to focus only on big, basic issues and to be rather dismissive of ideas that were more mundane and applied. His unswerving belief in the absoluteness of the entropy version of the second law, which he shared with few others, left him in a small minority within the scientific community. It also, curiously, led him to doubt the existence of atoms, and that was another irony, given how events turned out.

Like other scientists of his day, Planck was intrigued by why the universe seemed to run in only one direction, why time had an arrow, why nature was apparently irreversible and always running down. He was convinced that this cosmic one-way street could be understood on the basis of the absolute validity of the entropy law. But here he was out of step with most of his contemporaries. The last decade of the nineteenth century saw most physicists falling into line behind an interpretation of the second law that was the brainchild of the Austrian physicist Ludwig Boltzmann. It was while Boltzmann was coming of age and completing his studies at the University of Vienna that Clausius and Kelvin were hatching the second law, and Clausius was defining entropy and showing how the properties of gases could be explained in terms of large numbers of tiny particles dashing around and bumping into one another and the walls of their container (the so-called kinetic theory of gases). To these bold new ideas in the 1870s Boltzmann added a statistical flavor. Entropy, for example, he saw as a collective result of molecular motions. Given a huge number of molecules flying here and there, it's overwhelmingly likely that any organized starting arrangement will become more and more disorganized with time. Entropy will rise with almost, but not total, certainty. So, although the second law remains valid according to this view, it's valid only in a probabilistic sense.

Some people were upset by Boltzmann's theory because it just assumed from the outset, without any attempt at proof, that atoms and molecules exist. One of its biggest critics was Wilhelm Ostwald, the father of physical chemistry (and winner of the Nobel Prize in 1909), who wanted to rid physics of the notion of atoms and base it purely on energy—a quantity that could be observed. Like other logical positivists (people who accept only what can be observed directly and who discount speculation), Ostwald stubbornly refused to believe in anything he couldn't see or measure. (Boltzmann eventually killed himself because of depression brought on by such persistent attacks on his views.) Planck wasn't a logical positivist. Far from it: like Boltzmann, he was a realist who time and again attacked Ostwald and the other positivists for their insistence on pure experience. Yet he also rejected Boltzmann's statistical version of thermodynamics because it cast doubt on the absolute truth of his cherished second law. It was this rejection, based more on a physical rather than a philosophical argument, that led to him to question the reality of atoms. In fact, in as early as 1882, Planck decided that the atomic model of matter didn't jibe with the law of entropy. "There will be a fight between these two hypotheses that will take the life of one of them," he predicted. And he was pretty sure he knew which was going to lose out: "[I]n spite of the great successes of the atomistic theory in the past, we will finally have to give it up and to decide in favor of the assumption of continuous matter."

By the 1890s, Planck had mellowed a bit in his stance against atomism—he'd come to realize the power of the hypothesis even if he didn't like it—yet he remained adamantly opposed to Boltzmann's statistical theory. He was determined to prove that time's arrow and the irreversibility of the world stemmed not from the whim of molecular games of chance but from what he saw as the bedrock of the entropy law. And so, as the century drew to a close, Planck turned to a phenomenon that led him, really by accident, to change the face of physics.

While a student at Berlin from 1877 to 1878, Planck had been taught by Gustav Kirchoff who, among other things, laid down some rules about how electrical circuits work (now known as Kirchoff's laws) and studied the spectra of light given off by hot substances. In 1859, Kirchoff proved an important theorem about ideal objects that he called blackbodies. A blackbody is something that soaks up every scrap of energy that falls upon it and reflects nothing–hence its name. It's a slightly confusing name, however, because a blackbody isn't just a perfect absorber: it's a perfect emitter as well. In one form or another, a blackbody gives back out every bit of energy that it takes in. If it's hot enough to give off visible light then it won't be black at all. It might glow red, orange, or even white. Stars, for example, despite the obvious fact that they're not black (unless they're black holes), act very nearly as blackbodies; so, too, do furnaces and kilns because of their small openings that allow radiation to escape only after it's been absorbed and reemitted countless times by the interior walls. Kirchoff proved that the amount of energy a blackbody radiates from each square centimeter of its surface hinges on just two factors: the frequency of the radiation and the temperature of the blackbody. He challenged other physicists to figure out the exact nature of this dependency: What formula accurately tells how much energy a blackbody emits at a given temperature and frequency?

Experiments were carried out using an apparatus that behaved almost like a blackbody (a hot hollow cavity with a small opening), and equations were devised to try to match theory to observation. On the experimental side, the results showed that if you plotted the amount of radiation given off by a blackbody with frequency, it rose gently at low frequencies (long wavelengths), then climbed steeply to a peak, before falling away less precipitately on the high frequency (short wavelength) side. The peak drifted steadily to higher frequencies as the temperature of a blackbody rose, like the ridge of a barchan sand dune marching in the desert wind. For example, a warm blackbody might glow "brightest" in the (invisible) infrared band of the spectrum and be almost completely dark in the visible part of the spectrum, whereas a blackbody at several thousand degrees radiates the bulk of its energy at frequencies we can see. Scientists knew that this was how perfect blackbodies behaved because their laboratory data, based on apparatuses that were nearly perfect blackbodies, told them so. The sticking point was to find a formula, rooted in known physics, that matched these experimental curves across the whole frequency range. Planck believed that such a formula might provide the link between irreversibility and the absolute nature of entropy: his scientific holy grail.

Matters seemed to be moving in a promising direction when, in 1896, Wilhelm Wien, of the Physikalisch-Technische Reichsanstalt (PTR) in Berlin, gave one of the strongest replies to the Kirchoff challenge. "Wien's law" agreed well with the experimental data that had been gathered up to that point, and it drew the attention of Planck who, time and again, tried to reach Wien's formula using the second law of thermodynamics as a springboard. It wasn't that Planck didn't have faith in the formula that Wien had found. He did. But he wasn't interested in a law that was merely empirically correct, or an equation that had been tailored to fit experimental results. He wanted to build Wien's law up from pure theory and thereby, hopefully, justify the entropy law. In 1899, Planck thought he'd succeeded. By assuming that blackbody radiation is produced by lots of little oscillators like miniature antennae on the surface of the blackbody, he found a mathematical expression for the entropy of these oscillators from which Wien's law followed.

Then came a hammer blow. Several of Wien's colleagues at the PTR–Otto Lummer, Ernst Pringsheim, Ferdinand Kurkbaum, and Heinrich Rubens–did a series of careful tests that undermined the formula. By the autumn of 1900, it was clear that Wien's law broke down at lower frequencies–in the far infrared (waves longer than heat waves) and beyond. On that fateful afternoon of October 7, Herr Doktor Rubens and his wife visited the Planck home and, inevitably, the conversation turned to the latest results from the lab. Rubens gave Planck the bad news about Wien's law.

After his guests left, Planck set to thinking where the problem might lie. He knew how the blackbody formula, first sought by Kirchoff four decades earlier, had to look mathematically at the high-frequency end of the spectrum given that Wien's law seemed to work well in this region. And he knew, from the experimental results, how a blackbody was supposed to behave in the low-frequency regime. So, he took the step of putting these relationships together in the simplest possible way. It was no more than a guess–a "lucky intuition," as

Planck put it–but it turned out to be absolutely dead on. Between tea and supper, Planck had the formula in his hands that told how the energy of blackbody radiation is related to frequency. He let Rubens know by postcard the same evening and announced his formula to the world at a meeting of the German Physical Society on October 19.

One of the myths of physics, which even today is echoed time and again in books, both academic and popular, and in college courses, is that Planck's blackbody formula had something to do with what's called the "ultraviolet catastrophe." It didn't. This business of the ultraviolet catastrophe is a bit of a red herring (to mix colorful metaphors), worth mentioning here only to set the record straight. In June 1900, the eminent English physicist Lord Rayleigh (plain John Strutt before he became a baron) pointed out that if you assume something known as the equipartition of energy, which has to do with how energy is distributed among a bunch of molecules, then classical mechanics blows up in the face of blackbody radiation. The amount of energy a blackbody emits just shoots off the scale at the high-frequency end–utterly in conflict with the experimental data. Five years later, Rayleigh and his fellow countryman James Jeans came up with a formula, afterward known as the Rayleigh-Jeans law, that shows exactly how blackbody energy is tied to frequency *if* you buy into the equipartition of energy. The name "ultraviolet catastrophe," inspired by the hopelessly wrong prediction at high frequencies, wasn't coined until 1911 by the Austrian physicist Paul Ehrenfest. None of this had any bearing on Planck's blackbody work; Planck hadn't heard of Rayleigh's June 1900 comments when he came up with his new blackbody formula in October. In any case, it wouldn't have mattered: Planck didn't accept the equipartition theorem as fundamental. So the ultraviolet catastrophe, which sounds very dramatic and as if it were a turning point in physics, doesn't really play a part in the revolution that Planck ignited.

Planck had his formula in October 1900, and it was immediately hailed as a major breakthrough. But the forty-two-year-old theorist, methodical by nature and rigorous in his science, wasn't satisfied simply by having the right equation. He knew that his formula rested on little more than an inspired guess. It was vital to him that he be able to figure it out, as he had done with Wien's law–logically, systematically, from scratch. So began, as Planck recalled, "a few weeks of the most strenuous work of my life." To achieve his fundamental derivation, Planck had to make what was, for him, a major concession. He had to yield ground to some of the work that Boltzmann had done. At the same time, he wasn't prepared to give up his belief that the entropy law was absolute, so he reinterpreted Boltzmann's theory in his own nonprobabilistic way. This was a crucial step, and it led him to an equation that has since become known as the Boltzmann equation, which ties entropy to molecular disorder.*

We're all familiar with how, at the everyday level, things tend to get more disorganized over time. The contents of houses, especially of teenagers' rooms, become more and more randomized unless energy is injected from outside the system (parental involvement) to tidy them up. What Planck found was a precise relationship between entropy and the level of disorganization in the microscopic realm.

*The Boltzmann equation, derived by Max Planck in 1900, is $S = k \log W$, where S is entropy, W is molecular disorder, and k is a constant known as Boltzmann's constant.

To put a value on molecular disorder, Planck had to be able to add up the number of ways a given amount of energy can be spread among a set of blackbody oscillators; and it was at this juncture that he had his great insight. He brought in the idea of what he called energy elements–little snippets of energy into which the total energy of the blackbody had to be divided in order to make the formulation work. By the end of the year 1900, Planck had built his new radiation law from the ground up, having made the extraordinary assumption that energy comes in tiny, indivisible lumps. In the paper he presented to the German Physical Society on December 14, he talked about energy "as made up of a completely determinate number of finite parts" and introduced a new constant of nature, *h*, with the fantastically small value of about 6.6×10^{-27} erg second.* This constant, now known as Planck's constant, connects the size of a particular energy element to the frequency of the oscillators associated with that element.†

Something new and extraordinary had happened in physics, even if nobody immediately caught on to the fact. For the first time, someone had hinted that energy isn't continuous. It can't, as every scientist had blithely assumed up to that point, be traded in arbitrarily small amounts. Energy

*An erg is the energy or work involved in moving a force of one dyne a distance of one centimeter; a dyne is the force needed to accelerate a mass of one gram by one centimeter per second per second; an erg second–a unit of what physicists call "action"–is an erg expended over a period of one second.

†Planck's formula is $E = h\nu$, where E is the energy of an energy element or quantum, ν is its associated frequency, and h is Planck's constant.

comes in indivisible bits. Planck had shown that energy, like matter, can't be chopped up indefinitely. (The irony, of course, is that Planck still doubted the existence of atoms!) It's always transacted in tiny parcels, or *quanta.* And so Planck, who was anything but a maverick or an iconoclast, triggered the transformation of our view of nature and the birth of quantum theory.

It was to be a slow delivery. Physicists, especially Planck (the "reluctant revolutionary" as one historian called him), didn't quite know what to make of this bizarre suggestion of the graininess of energy. In truth, compared with all the attention given to the new radiation law, this weird quantization business at its heart was pretty much overlooked. Planck certainly didn't pay it much heed. He said he was driven to it in an "act of despair" and that "a theoretical interpretation *had* to be found at any price." To him, it was hardly more than a mathematical trick, a theorist's sleight of hand. As he explained in a letter written in 1931, the introduction of energy quanta was "a purely formal assumption and I really did not give it much thought except that no matter what the cost, I must bring about a positive end." Far more significant to him than the strange quantum discreteness (whatever it meant) was the impressive accuracy of his new radiation law and the new basic constant it contained. This lack of interest in the strange energy elements has led some historians to question whether Planck really ought to be considered the founder of quantum theory. Certainly, he didn't see his work at the time as representing any kind of threat to classical mechanics or electrodynamics. On the other hand, he did win the 1918 Nobel Prize for Physics for his "discovery of energy quanta." Perhaps it would be best to say that Planck lit the fuse and then withdrew. At any rate, the reality of energy quanta was definitely put on a firm footing a few years later, in 1905–by the greatest genius of the age, Albert Einstein.

Mention Einstein and the first thing that springs to mind is the theory of relativity, that other extraordinary supernova that burst upon twentieth-century physics. Yet, incredibly, Einstein never won a Nobel Prize for relativity. His one Nobel medal (he surely should have gotten at least two), awarded in 1921 and presented in 1922, was for his pioneering work in quantum theory. If Planck hadn't fathered quantum theory, that role may well have fallen to Einstein. As it was, Einstein was the first person to take the physical implications of Planck's work seriously. The turning point came when he saw how Planck's idea of energy quanta could be used to account for some puzzling facts that had emerged about a phenomenon known as the photoelectric effect.

In 1887, Heinrich Hertz became the first person to observe the photoelectric effect during his experiments that confirmed Maxwell's theory of electromagnetism. Hertz found that by shining ultraviolet light onto metal electrodes, he could lower the voltage needed to make sparks hop between the electrodes. The light obviously had some electrical effect, but Hertz stopped short of speculating what that might be. "I confine myself at present," he said, "to communicating the results obtained, without attempting any theory respecting the manner in which the observed phenomena are brought about."

In 1899, the English physicist J. J. Thomson offered an important clue toward understanding the photoelectric effect. Thomson showed that ultraviolet light, falling onto a metal surface, triggered the emission of electrons.* These were tiny, charged particles whose existence Thomson had demonstrated a couple of years earlier and which he believed were the only material components of atoms. The photoelectric effect, it seemed to physicists at the time, must come about because electrons inside the atoms in a metal's surface were shaken and made to vibrate by the oscillating electric field of light waves falling on the metal. Some of the electrons would be shaken so hard, the theory went, that eventually they'd be tossed out altogether.

In 1902, Philipp Lenard, who'd earlier been an assistant to Hertz at the University of Bonn, made the first quantitative measurements of the photoelectric effect. He used a bright carbon arc light to study how the energy of the emitted photoelectrons varied with the intensity of the light and, by separating out individual colors, with the frequency of light. Increasing the frequency of light by selecting light from the bluer end of the spectrum caused the ejected electrons to be more energetic on average, just as predicted—because, it was assumed, these electrons had been made to vibrate faster. Increasing the intensity of light (by moving the carbon arc closer to the metal surface) caused more electrons

*The name *electron* was actually coined before the discovery of the particle. It was first used in 1891 by the Irish physicist G. Johnstone Stoney to denote the unit of charge found in experiments that passed electric current through chemicals. Thomson initially referred to his newfound particles simply as *corpuscles*. Another Irish physicist, George FitzGerald, suggested in 1897 that Thomson's corpuscles were really *free electrons*, though he disagreed with Thomson that electrons also occurred inside atoms.

to be thrown out, also as expected. On the other hand, increasing the intensity had no effect at all on the average amount of energy that each ejected electron carried away. That came as a real shock. If, as physicists believed, the photoelectric effect followed from an interaction between electrons and electromagnetic waves, then intensifying the radiation ought to shake the electrons in the metal surface harder and therefore shoot them out with more energy. It was a mystery why this didn't happen.

Several years went by before Lenard's observation on the photoelectric effect and Planck's strange but neglected theory of the quantum, both puzzling in themselves, were seen as arrows pointing toward a common solution. Looking back now, it seems clear enough, but it took the genius of Einstein to apply quantization, not to blackbody oscillators as Planck had done in a desperate effort to patch up classical theory, but to the actual radiation that's emitted or absorbed. *Light itself is quantized*, Einstein realized. All the light of a particular frequency comes in little bullets of the same energy, equal to the frequency multiplied by Planck's constant, and that's the key to understanding the photoelectric effect. An incoming light quantum smashes into an electron on the surface of a metal and gives up all of its energy to the electron. A certain amount of energy, called the work function, is needed simply to overcome the force of attraction between the electron and the metal in order to set the electron free; so there can't be any photoelectric effect unless this threshold is reached. Any energy left over from the exchange, above and beyond the work function, appears as kinetic energy (energy of motion) of the ejected electron. Increasing the intensity of radiation–the number of light quanta per unit area–has no effect on the energy of individual electrons because each electron is thrown out by one and only one parcel of light. Increasing the frequency of radiation, on the other hand, means that each light bullet packs a bigger wallop, which results in a more energetic photoelectron.

The fact that sixteen years went by before Einstein won a Nobel Prize for his groundbreaking work on the photoelectric effect reflects how long it took the scientific world to accept that radiant energy is quantized. That may seem like an age, but the idea that energy, including light, is granular ran counter to everything that physicists had been taught for several generations: matter is made of particles; energy is continuous and tradable in arbitrarily small amounts; light consists of waves; matter and light don't intermingle. These rules had been the mantras of physics for much of the nineteenth century and were now being overturned.

There was also the issue of experimental proof. It took a decade or so for the details of Einstein's photoelectric theory to be thoroughly tested and verified in the lab. The actual observation that the kinetic energy of electrons kicked out by the photoelectric effect is tied to the frequency of incoming light in exactly the way Einstein prescribed was finally made in 1916 by the American physicist Robert Millikan. Millikan had, in fact, long been expecting to prove Einstein wrong and thereby to uphold the wave theory of light. Instead he wound up giving powerful support to the particle theory and measuring Planck's constant to within 5 percent of its currently accepted value. Ironically, he won the Nobel Prize in 1923 for a superb series of experiments that dashed what earlier had been his greatest scientific hope.

We talk about the quantum revolution–but it wasn't an overnight affair, this overthrow of the old worldview of matter and energy in favor of a new one. It was more than two decades after Planck's first inkling of the existence of quanta when quantum theory was fully accepted and acknowledged as the reigning paradigm of the microcosmos. For the first part of this interregnum, Einstein was at the cutting edge of developments. Following his seminal 1905 photoelectric paper, he worked on meshing Planck's notion of the quantum with other areas of physics. For instance, he showed that some anomalies having to do with how much heat substances have to absorb to raise their temperature by a certain amount are best explained if the energy of vibration of atoms is assumed to be quantized. This early quantum pioneering by Einstein now seems almost entirely overshadowed by his work on relativity, but it was instrumental at the time in persuading scientists of the validity of quantum theory when applied to matter.

His views on the quantum nature of electromagnetic radiation proved a harder sell. Yet he insisted that the way ahead had to lie with some acceptance of light's particlelike behavior. In 1909 he wrote: "It is my opinion that the next phase in the development of theoretical physics will bring us a theory of light that can be interpreted as a kind of fusion of the wave and emission theory." In 1911, at the first Solvay Congress (an annual meeting of the world's top physicists) he was more forceful: "I insist on the provisional character of this concept, which does not seem reconcilable with the experimentally verified consequences of the wave theory." That apparent irreconcilability was a major stumbling block for all scientists. What kind of madness was it to argue that light could be both a particle and a wave?

Experimentalists railed at the prospect of what Einstein's equation of the photoelectric effect implied. Robert Millikan, the very man who showed that the equation really did work, would have nothing to do with its physical interpretation. In 1915, Millikan wrote: "The semicorpuscular theory by which Einstein arrived at his equation seems at present wholly untenable." Three years later, Ernest Rutherford, the great New Zealand physicist who probed the structure of the atom, said there appeared to be "no physical connection" between energy and frequency in Einstein's hypothesis about light quanta. It didn't seem to make sense that a particle could have a frequency, or that a wave could act as if it were made of energetic particles. The two concepts seemed to rule each other out.

Between 1911 and 1916, Einstein took a sabbatical from his quantum work to attend to another little problem–the general theory of relativity, which transformed our ideas on gravity. Upon his return to the physics of the very small, he quickly grasped a link between quantum theory and relativity that convinced him of the reality of the particle aspect of light. In earlier work, Einstein had treated each quantum of radiation as if it had a momentum equal to the energy of the quantum divided by the velocity of light. By making this assumption he was able to explain how momentum is transferred from radiation to matter–in other words, how atoms and molecules are buffeted when they absorb radiation. Although this buffeting was much too small to be seen directly, it had effects on properties, such as the pressure of a gas, that *could* be measured. These measurements fit with the formula for quantized momentum. Einstein now realized, in coming back to his quantum studies, that exactly the same expression for the momentum of a light quantum fell straight out of a basic equation in relativity theory.* This link between relativity and the earlier assumption about the momentum of a radiation quantum clinched the case for light particles in Einstein's mind. In 1917, he may have been the only major scientist alive who believed that light had a genuine particle aspect. But the fact that his theory now insisted that whenever these supposed light quanta interacted with particles of ordinary matter a definite, predictable amount of momentum should be transferred paved the way for experimental tests. Six years later, the particle nature of light had been found to be virtually beyond dispute.

At the heart of the lab work that ultimately proved the reality of radiation quanta was the American physicist Arthur Compton. In his early days at Princeton, Compton devised an elegant way of demonstrating Earth's rotation, but he soon launched into a series of studies involving X-rays that climaxed in the final victory of quantum physics over the old world order. In his mid-twenties Compton hatched a theory of the intensity of X-ray reflection from crystals that provided a powerful tool for studying the crystallographic arrangement of electrons and atoms in a substance. In 1918 he began a study of X-ray scattering that led inevitably to the question of what happens when X-rays interact with electrons. The

*Einstein took the momentum of a quantum to be $h\nu/c$, where $h\nu$ is the energy of the quantum and c is the velocity of light. In the special theory of relativity, published in 1905, the energy (E), momentum (p), and rest mass (m) of a particle are tied together by the equation: $E^2 = m^2c^4 + p^2c^2$. Rest mass, as the name suggests, is the mass of a particle when it's standing still relative to an observer. Taking the hypothetical rest mass of a light quantum to be zero (which has to be the case to avoid the mass of a light particle traveling with velocity c being infinite), this equation reduces to $E = pc$, so that $p = E/c$, exactly as before.

key breakthrough came in 1922 and was published the following year. Compton found that when X-rays scatter from free electrons (electrons not tightly bound inside atoms) the wavelength of the X-rays increases. He explained this effect, now known as the Compton effect, in terms of radiation quanta colliding with electrons, one quantum per electron, and giving up some of their energy (or momenta) in the process. Energy lost translated to frequency decrease, or wavelength increase, according to the Planck formula. A further boost for this interpretation came from a device invented by Charles Wilson. Inspired by the wonderful cloud effects he'd seen from the peak of Ben Nevis, in his native Scotland, Wilson built a vessel in which he could create miniature artificial clouds. This cloud chamber proved invaluable for studying the behavior of charged particles, since water droplets condensing in the wake of a moving ion or electron left a visible trail. Wilson's cloud chamber revealed the paths of the recoil electrons in the Compton effect, showing clearly that the electrons moved as if struck by other particles–X-ray quanta–which, being uncharged, left no tracks. Final proof that the Compton effect really was due to individual X-ray quanta scattering off electrons came in 1927 from experiments based on the so-called coincidence method, developed by Walter Bothe. These experiments showed that individual scattered X-ray quanta and recoil electrons appear at the same instant, thereby laying to rest some arguments that had been voiced to try and reconcile quantum views with the continuous waves of electromagnetic theory. To complete the triumph of the particle picture of light, the American physical chemist Gilbert Lewis coined the name "photon" in 1926, and the fifth Solvay Congress convened the following year under the title "Electrons and Photons."

Doubt had evaporated: light could manifest itself as particles. But there was equally no doubt that, at other times, it could appear as waves, and that didn't seem to make any sense at all. As Einstein said in 1924, "There are . . . now two theories of light, both indispensable . . . without any logical connection."

By the mid-1920s, it was obvious–from the photoelectric effect, from the Compton effect, and in other ways–that when light interacts with matter it does so as if it were made of tiny bullets of energy called photons. The rest of the time it goes about as if it were smeared out in the form of a wave. Apparently, light has an identity crisis, and nowhere was that crisis more evident than in an updated version of an experiment carried out long before quantum theory came on the scene.

Thomas Young's double-slit experiment, dating back to the early nineteenth century, offers the clearest, most unambiguous proof of the wavelike personality of light. On the screen at the rear of the apparatus appears a series of alternating bright and dark bands. Two waves from a common source, one rippling out from each slit, combine, and the stripes on the screen speak unarguably of the adding and canceling out of wave crests and troughs.

What happens now if we dim the light source? A standard 60-watt light bulb puts out roughly 150 million trillion photons per second. This vast number underscores why quantum effects, which expose the discreteness of energy, go unnoticed at the everyday level: the individual energy transactions involved are fantastically small. If we want to pursue the question of how light *really* behaves, on a tiny scale, we have to turn the lamp in Young's experiment down–way down. This was first done in 1909 by the English physicist and engineer Geoffrey Taylor, who was later knighted for his work in aeronautics.[47] Shortly after his graduation from Cambridge, Taylor set up a version of Young's experiment using a light source so feeble that it was equivalent to "a candle burning at a distance slightly exceeding a mile." Even at this level of illumination, the interference pattern showed up. The dribble of light passing through the apparatus continued to behave in a wavelike manner.

What if the light source were turned down even further? What if it were dimmed so much that it effectively spat out single photons? There was no way to arrange for this to happen in the early twentieth century; the technology needed just wasn't at hand. Fashioning a light source that emits only one photon at a time isn't as simple as turning on a faucet so that water comes out drip by drip (after all, each water droplet contains many trillions of atoms, each of which is more substantial than a photon). Consequently, those involved in the formative phase of quantum physics, like those grappling with early relativity theory, had to rely on *gedanken*–thought experiments–to test their ideas. *If* Young's experiment could be done using a light source that fired out individual photons, what would be seen on the screen? The only answer that squared both with experiments that *had* been carried out and with the emerging principles of quantum theory is that the interference pattern would build up, one point at a time. This ought to happen even if there was no more than a single photon passing through the apparatus at any given moment. As the English theoretical physicist Paul Dirac put it: "Each photon then interferes only with itself."

Today double-slit experiments with single photons are routinely set up as demonstrations for undergraduates. An arrangement used at Harvard, for example, employs a helium-neon laser as a light source, two rotatable Polaroid filters to cut the intensity down to barely visible, and a pinhole that is 26 microns (millionths of a centimeter) in diameter at the front end of a PVC pipe. Further down the pipe is a slide with slits, each 0.04 millimeter wide, set 0.25 millimeter apart. Light from the double-slit then falls onto a sensitive video camera, which produces an image on a screen in which individual flashes, corresponding to single photons, can be seen appearing. With the detection equipment in storage mode, the single flashes of light are captured live, and the characteristic double-slit interference pattern can be watched building up in real time. The familiar bright and dark bands, which cry out for a wave interpretation, emerge like a pointilist painting from the specks that are obviously the marks of individual colliding particles.

Something very, very strange is going on here. In the single-photon, double-slit experiment, each photon starts and ends its journey as a particle. Yet in between it behaves as if it were a wave that had passed through both slits, because that's the only way to account for the interference pattern that forms over time. During its flight from source to detector, the photon acts in a way that defies not only commonsense but all of physics as understood before Planck and Einstein.

You might say, let's keep closer tabs on each photon in the experiment. If it's a particle–a single, pointlike entity–it can't *really* go through both slits at once, any more than a person can simultaneously walk through two doorways. What happens if we put a detector on one of the slits to tell us whether the photon goes through that slit or the other? This is easy to arrange and, sure enough, the photon is forced to give itself up. By posting a sentry at one of the slits, we learn which slit each photon passes through. But in gaining this knowledge, we lose something else: the interference pattern. Flushed into the open, compelled in midflight to reveal its whereabouts, the photon abruptly abandons its wavelike behavior and acts purely and simply like a miniature bullet bound on a straight-line trajectory. Somehow the existence of the interference pattern is tied to a lack of knowledge as to which slit the photon actually went through. If we don't ask where the photon is, it behaves like a wave; if we insist upon knowing, it behaves like a particle. In classical physics such a situation would be unthinkable, outrageous. Yet there it is: the act of observing light makes its wave nature instantly collapse and its particle aspect become manifest at a specific point in space and time. It's almost as if a photon knows when it's being watched and alters its behavior accordingly. Evidently, an enigma lies at the heart of the quantum world that, like a Zen koan, resists a solution in familiar, everyday terms. Yet we have only begun to touch upon the mystery. This astonishing wave-particle duality isn't confined to photons. The strange goings-on of the double-slit experiment are merely a prelude to more quantum weirdness that enables the fantasy of teleportation to come true.

2

Ghosts in the Material World

Light is ethereal; matter is solid. Light is mercurial and lacks substance; matter is (reasonably) permanent. You know where you are with matter because you *are* matter. It's the stuff of your body and brain and of all the substances and objects in the universe around you. Perhaps we shouldn't be surprised by some oddities in the case of light and other forms of radiation, which are, after all, such slippery, intangible things. But there can be none of these quantum shenanigans with matter; can there?

• • •

The first signs that the material world is disturbingly different from what it seems came in 1909. Two researchers in Ernest Rutherford's lab at the University of Manchester, Hans Geiger and Ernest Marsden, fired a beam of alpha particles at a thin metal foil. Alpha particles had been identified and named (they were called "alpha rays" to begin with) a decade earlier by Rutherford, as one of the types of radiation given off by radioactive elements such as uranium. Rutherford reasoned that as alpha particles are fast-moving and positively charged (they're now known to be high-speed helium nuclei), they'd serve as a good probe of the atomic structure of matter. According to one popular theory at the time, championed by J. J. Thomson, atoms were built like currant buns–with electrons (particles carrying negative charge) for currants and a smeared-out positive charge for the rest of the bun to keep the whole thing electrically neutral. If the currant bun model were right, every one of the alpha particles ought to zip straight through the thin foil in Geiger and Marsden's experiment suffering, at most, minor deflections because of the weak influence of the spread-out positive charge (the effect of the electrons being negligible). What Geiger and Marsden actually found was stunning. Most of the alpha particles did indeed travel straight through the foil with little or no deviation, but a small fraction (about one in ten thousand) rebounded, ending up on the same side of the foil as the incoming beam. A few were returned almost along the same tracks as they went in. Rutherford described hearing of these rebounds as the most incredible event of his life. It was, he said, "as if you fired a 15-inch shell at a piece of tissue paper and it came back and hit you." Such huge deflections could mean only one thing: some of the alpha particles had run into massive concentrations of positive charge and, since like charges repel, had been hurled straight back by them.

Thomson's currant bun model of the atom, in which positive charge was spread thinly over the whole atom, hadn't a hope of explaining the results. Instead, in 1911, Rutherford cooked up a new model of the atom in which all of the positive charge is crammed inside a tiny, massive nucleus about ten thousand times smaller than the atom as a whole. That's equivalent to a marble in the middle of a football stadium. The much lighter electrons, he assumed, lay well outside the nucleus. To the shock and amazement of everyone, the atoms of which planets, people, pianos, and everything else are made consisted almost entirely of empty space.

Rutherford's nuclear model of the atom was a huge step forward in understanding nature at the ultrasmall scale. But even as it closed the casebook on the alpha particle experiment, it threw open another one. Since the nucleus and its retinue of electrons are oppositely charged, and therefore attract one another, there didn't appear to be anything to stop the electrons from being pulled immediately into the nucleus. Throughout the universe, atomic matter ought to implode in the wink of an eye. Rutherford countered by saying that the atom was like a miniature solar system: the electrons circled the nucleus in wide orbits just as planets orbit the sun. This is the picture of atoms that most of us still carry around in our heads. It's an appealing, easy-to-grasp image–one that's inspired many a logo of the atomic age. Yet theorists were well aware of its shortcomings right from the start.

The fatal flaw in Rutherford's model is that it contains charges that are *accelerating*. The charges are on the electrons and the acceleration is due to the electrons always changing direction as they move around their orbits. (Things accelerate when they change speed and/or direction.) Since Maxwell's time, scientists had known that accelerating charges radiate energy. What was to stop the orbiting electrons in Rutherford's atom from quickly (in fact, in about one hundred-millionth of a second) losing all their energy and spiraling into the nucleus?

The answer came from a young Dane, Niels Bohr, who joined the team at Manchester for a six-month spell in 1912, shortly after Rutherford went public with his new vision of the atom. Bohr played a hunch. He knew about Planck's quantum. He knew there was no way to prevent an electron inside an atom from plummeting into the nucleus if it could give off energy continuously. So he said simply that electrons inside atoms *can't* radiate continuously. They can only radiate in lumps, and these lumps are the same as Planck's quanta. For a given type of atom, say hydrogen, there's a limited number of stable orbits that an electron can occupy. Each of these orbits corresponds to a whole multiple of the basic quantum.* As long as an electron is in one of these orbits, its energy, contrary to whatever classical physics might say, stays the same. If it jumps from an outer (higher energy) orbit to an inner (lower energy) orbit, the energy difference between the two is given off as a quantum of light. Once the electron reaches the lowest energy orbit, it can't fall any farther and is safe from the clutches of the nucleus.

Bohr's next problem was how to test this startling idea. He didn't have a microscope powerful enough to peek inside an atom and look for the special, energy-conserving orbits.

*To be exact, the allowed orbits in the Bohr atom are ones for which the angular momentum of the electron is a whole number multiple of $h/2\pi$, where h is Planck's constant.

He did have something almost as good, however: the spectrum of the atom. Each element has its own spectrum, as unique as a fingerprint, marked by bright lines at specific wavelengths. Bohr's great insight was to realize that each of these lines corresponded to a particular energy jump within an atom of that element. He calculated the stable orbits of hydrogen (the simplest atom, with just one electron) and, in 1913, showed that some of the possible transfers an electron can make between these orbits exactly matched the known lines in the visible, ultraviolet, and infrared parts of the hydrogen spectrum. A bright red line, for example, corresponded to an electron jump from the third orbit to the second; a vivid blue-green line fell at the precise wavelength expected of an energy quantum emitted by an electron jumping from the fourth orbit to the second. To round out the success of Bohr's model, other spectral lines of hydrogen that it predicted in the X-ray region, never before seen, were quickly confirmed by Harold Moseley, another member of Rutherford's team.

Just as Planck put the quantum into the theory of blackbody radiation (for no other reason than the fact that it worked), Bohr quantized the atom to save Rutherford's miniature solar system theory from collapse. It was an important step, and it won Bohr the 1922 Nobel Prize for Physics. But Bohr's atom was still a hodgepodge of classical and quantum ideas, and, despite some ingenious fine tuning over time, it continued to have faults. It predicted far more spectral lines than were actually seen, so that these additional lines had to be arbitrarily declared "forbidden." To fit other observations, new quantum numbers describing unknown atomic properties had to be conjured up out of thin air. The Bohr atom was a valuable stopgap, and, crucially, it showed that no theory of atomic processes could work unless it embraced quantum ideas. Yet it lacked one vital ingredient that was needed for the full flowering of this strange new science: the element of chance.

The missing piece of the quantum puzzle first surfaced in the early 1900s when Rutherford was working in Canada with a colleague, the chemist Frederick Soddy. Rutherford had already shown that radioactive elements such as uranium and radium give off three different types of radiation, which he called alpha, beta, and gamma rays. (The first two were found in 1898, gamma rays two years later.) Now, together with Soddy, he made the far more revolutionary discovery that in emitting these rays the elements involved changed, or decayed, by a cascade of transformations, into other elements. This was the ancient dream of alchemists: to transmute one element into another. But the way in which it came true was very puzzling. Radioactive decay didn't seem to be prompted or influenced by any outside factor; it just happened. A radioactive substance would decay spontaneously and at a rate sublimely indifferent to whether it was disturbed or left alone, heated or cooled, pressurized or put in a vacuum. Rutherford and Soddy found that each radioactive species had its own characteristic half-life–the time needed for half the original number of atoms to decay. Radium-226, for instance, which is the most stable form of this extremely unstable element, has a half-life of 1,600 years, so that after this period 50 percent of the atoms in the starting sample of radium will have transmuted into atoms of the deadly gas radon.

In bulk, radioactive atoms follow a predictable pattern of

decay, but the behavior of an individual atom is totally unpredictable. It might decay in a split second, in a century, or in a million years. Compare this with a human population. For a large group of people, there's an average lifespan. In the United States in 2001, for example, the average life expectancy was 74.4 years for men and 78.8 for women. These are the figures you'd be best to put your money on if a newborn baby were picked out of the population at random. The actual lifespan of an individual depends on all kinds of factors, however, some internal, such as genetic makeup and state of health, and others external, such as the environment (exposure to carcinogens, for example) and accidents. The lifetime of a radioactive nucleus can also be totally different from the average (as measured by the half-life). But whereas with individual people we can always find a reason–if we look closely enough–why they died when they did, radioactive decay appears to be genuinely and totally random.

Rutherford and Soddy didn't know what was at the bottom of radioactive decay but they realized they had a statistical situation on their hands. To deal with it, they turned to the kind of actuarial techniques used by life insurance companies in reckoning the likelihood that an individual client will be dead within so many years. Through their statistical theory of radioactive decay, Rutherford and Soddy could predict what proportion of a sample of a particular radioactive substance would decay within a given amount of time. But the fate of a specific atom seemed to be in the hands of the gods. Chance had entered the physics of the atom, and the most profound, and profoundly puzzling, property of quantum systems had made its presence felt for the first time.

It wouldn't long escape the attention of Einstein. In 1916,

with general relativity fresh under his belt, Einstein turned his attention back to the affairs of the atomic world. It occurred to him that the transition an atom makes in going from a state of higher energy (known as an excited state) to one of lower energy has a lot in common with the decay of a radioactive atom. As with radioactive decay, the jump happens without prompting and apparently at random. Einstein, too, approached this problem from the standpoint of statistics. He combined methods used by Boltzmann in the nineteenth century for describing the behavior of collections of atoms with the actuarial tables applied to radioactivity. This allowed him to figure out the probability that an electron will jump from an energy state described by a higher quantum number to a state with a lower quantum number. Soon after, Bohr applied Einstein's new results to his atomic model and was able to show, on the basis of transition probabilities, why some spectral lines are much stronger than others.

But a deeper mystery remained. Why did things at the atomic and subatomic levels behave in this capricious way at all? Why did a particular nucleus decay, or a particular electron make an energy jump when it did, and not at some other time? In the pioneering days of quantum theory, physicists assumed that behind the apparent whimsy of such events were underlying causes and that it was just a matter of time before someone found the hidden deterministic processes at work.

The period from 1900 to 1926 was one of gradual change in physics during which ideas about the quantization of energy seeped into the scientific consciousness and were tested better and better by observation. It was the heyday of what's now called the "old quantum theory" during which scientists such as Planck, Bohr, and Einstein, who had been raised on a diet of classical physics, struggled to graft quantum concepts onto the comfortable, commonsense physics of the nineteenth century. The hope of these quantum frontiersmen was that, in time, the quirky behavior of atoms and radiation could be reconciled with Newtonian mechanics and Maxwellian electromagnetism–that atoms would ultimately take their place in the vast, predictable machinery of the great cosmic clockwork. But it was not to be. The quantum world was far stranger than even its most adventurous early explorers had guessed or feared.

In July 1925, Bohr sounded the death knell for the old quantum theory: "One must be prepared for the fact that the required generalization of the classical electrodynamical theory demands a profound revolution in the concepts on which the description of nature has until now been founded." There could be no more compromises, no more attempts to fit quantum concepts within the framework of classical physics. Nothing short of a revolution would do. In fact, it had already begun.

Just as Planck sowed the seeds of the old quantum theory in the course of an evening, his compatriot Werner Heisenberg hatched the new quantum theory, which became known as quantum mechanics,* in the space of a few hours a quarter of a century later. Heisenberg was born the year after Planck's discovery of the quantum, so he was never held back by any ingrained prejudice against the fantastic new

*The phrase "quantum mechanics," analogous to Newtonian mechanics, was first used in the title of a paper, "Zur Quantummechanik," by Max Born in 1924.

ideas about the atom. What's more, his interest in Greek philosophy, especially that of Plato and the atomists (ancient thinkers who believed in the discreteness of matter), led him to think of atoms in strictly abstract terms, not as things that needed to be pictured or imagined. He'd been tempted to go into pure mathematics, but then, as a student, met Bohr and through him became engrossed in atomic theory. In 1924, after wrapping up his doctoral work at Göttingen under Arnold Sommerfeld, another of the great early quantum pioneers, Heisenberg went to join Bohr in the newly founded Niels Bohr Institute in Copenhagen, before returning to become assistant to Max Born in the physics department at Göttingen. The young German had taken up the problem of the spectral lines of hydrogen, details of which continued to elude the Bohr model even with all the fixes that had been made to it since its inception.

On June 7, 1925, Heisenberg left for the rocky North Sea island of Helgoland to convalesce after a bad attack of hay fever. But he hardly rested at all, what with mountain climbing, memorizing poems by Goethe, and, especially, poring over the problem of the hydrogen atom. His obsession paid off. "It was about three o'clock at night when the final result of the calculation lay before me," he recalled in his autobiographical *Physics and Beyond.* "At first I was deeply shaken. I was so excited that I could not think of sleep. So I left the house and awaited the sunrise on the top of a rock."

Upon returning to Göttingen, Heisenberg sent a copy of his results to his close colleague Wolfgang Pauli, a fellow rising star in the quantum firmament, with a note saying: "Everything is still vague and unclear to me, but it seems as if the electrons will no more move on orbits." Absent from Heisenberg's scheme was any hint of miniature solar systems, or of material components at all. Instead, the atom and its energy jumps were reduced to an array of numbers and to rules by which these numbers could be manipulated. Heisenberg quickly wrote a paper on his findings and, with some trepidation, gave a copy to Born, who admitted to being at first completely astonished by the strangeness of the calculations put forward. Then, one morning in July, Born suddenly realized where he'd seen that type of calculation before. It was in a branch of mathematics developed fifty years earlier, long before there was any practical use for it, to handle arrays of numbers called matrices. As soon as Born saw the connection, he and one of his students, Pascual Jordan, began recasting Heisenberg's results in formal matrix language. When Heisenberg was up to speed with the techniques of handling matrices, he joined in the collaboration. So, within the space of a few months in the middle of 1925, the first complete mathematical portrait of the new quantum mechanics–the form known as matrix mechanics–was set down.

In classical physics, the variables that tell how something moves are just ordinary numbers. For instance, a particle might be 8 meters from a fixed point, so we could say that its position (call it q) was 8; its momentum (call it p), equal to mass times velocity, might come out to be, say, 5, in some appropriate units. The order in which simple numbers like these are multiplied doesn't matter–$8 \times 5 = 5 \times 8$; the numbers are said to obey the commutative law of multiplication. What alarmed Heisenberg when he first began calculating with his arrays is that the order in which they were multiplied *did* matter. It worried Born, too, until he remembered that one of the most striking features of matrix multiplication is that it's noncommutative. In the matrix mechanics formulation of quantum theory that followed, physical variables such as position q and momentum p are no longer represented by simple numbers but by matrices, and it's no longer the case that p times q has to equal q times p. In fact, Born and Jordan quickly figured out that the quantum world equivalents of position and momentum satisfied a very peculiar relationship: the difference between p times q and q times p is equal to a number that involves Planck's constant. To be precise, $pq - qp = ih/2\pi$, where h is Planck's constant and i is the square root of minus one. So basic and profound is this formula, and so radically unlike anything found in Newtonian mechanics, that Born, Jordan, and Heisenberg referred to it in their first joint paper as the "fundamental quantum-mechanical relation." Years later it was etched on Born's gravestone. The great question was: what did it mean?

Even as matrix mechanics took root in Göttingen, another very different-looking version of new quantum theory was emerging elsewhere. Its author was the Austrian Erwin Schrödinger and its birthplace was Zürich, where since 1921 Schrödinger had been professor of physics. Both man and place bucked the trend in quantum developments at this time. Whereas Heisenberg, Pauli, Jordan, and others who were breaking new ground in the field were young turks born in the post-Planck era, Schrödinger was thirty-nine at the time of his greatest discovery and Zürich was a very conservative place, not at all a hotbed of revolutionary thinking. Yet conservatism was the hallmark of this alternative scheme of quantum mechanics–or so it seemed at the outset. While the matrix approach went out of its way to be abstract, Schrödinger's treatment was overtly pictorial and an unabashed attempt to see in this strange new quantum landscape something familiar and accessible.

Schrödinger had heard nothing of the matrix developments in Germany. His inspiration came instead from the work of Louis de Broglie, the second son of a duke, who submitted a remarkable doctoral thesis at the Sorbonne in 1924. De Broglie was himself no youngster in this brave new quantum world. The thirty-two-year-old had started late in physics, having switched from a planned career in the civil service, his scientific ambitions further put on hold by World War I during which he served as a telegraph operator atop the Eiffel Tower. After the war, Louis worked with his older brother Maurice, whose experiments in his laboratory at the family mansion helped bolster the case for Einstein's particle description of light. Louis was more drawn to the theoretical side and to questions about the deep nature of things. It was while thinking about Einstein's new ideas on light that he reached his astounding conclusion, which in retrospect (like many great insights) makes perfect sense: if waves can behave like particles, then particles can behave like waves.

De Broglie first published his claim, that wave-particle duality should be extended beyond light to include matter, in three short papers published in 1923. It was in his doctoral thesis the following year, however, that he packaged his ideas in the form that triggered a major advance in physics. De Broglie pointed out that the failure of experiments to decide whether light was *really* a particle or a wave suggested that these properties always go hand in hand–the one can't exist without the other. Furthermore, the fact that electrons in atoms occupy only orbits corresponding to whole numbers of quanta hinted at a wavelike nature. "The only phenomena involving integers in Physics," he later wrote, "were those of interference and of normal modes of vibration. This . . . suggested to me . . . that electrons too could not be regarded simply as corpuscles, but that periodicity must be assigned to them." De Broglie proposed that the allowed electron orbits were those in which an exact number of whole standing waves (like the vibrations of a plucked string), associated with the electrons, could fit: one complete up-and-down cycle in the lowest orbit, two in the second, and so on. Starting from Einstein's formulas for light quanta, he reasoned that the wavelength of the wave associated with an electron (whether inside an atom or out) is equal to Planck's constant divided by the momentum, and suggested looking for the diffraction of electrons as evidence of their wavelike behavior.

Einstein received a copy of de Broglie's thesis from one of the examiners and was impressed by it. Shortly after, he wrote: "I believe that it involves more than merely an analogy." Encouraged by the great man's endorsement, Schrödinger seized upon the notion of de Broglie's waves to build a complete mathematical wave model of the atom. After some setbacks, he succeeded, publishing his results in a series of papers in 1926. At the heart of this new conception was an equation similar in form to those that describe waves in the everyday world, such as water waves and sound waves. It seemed at first that atomic physics had been brought back from the brink of paradox and weirdness by a theory that painted, as in prequantum days, reassuring pictures of orbits, waves, and predictable events. Because of this familiar feel, and because the mathematics was a lot easier to understand than that of its matrix rival, wave mechanics quickly became the favored model of quantum theory. It was almost like old times. As in the days of Newton and Huygens, you could take your pick between a particle approach, which matrix mechanics leaned toward even though it dealt in abstract quantities, and Schrödinger's scheme, which was strongly wave-biased. In the background was still that nagging issue of duality–the fact that experiments could be set up to detect either aspect. But a brief period ensued when men like Schrödinger thought they could see a new realism emerging in the physics of the quantum.

De Broglie fostered the belief that particles–electrons, in particular–might somehow be *associated* with waves. He talked about each electron having a "pilot wave" that guided its motion. In 1927, the American physicists Clinton Davisson and Lester Germer observed the diffraction of electron beams from a nickel crystal, demonstrating the wavelike properties of particles of matter for the first time.*

But the illusion that the wave model somehow offered a superior picture of the atom, or that the waves in question were as real and physical as, say, vibrations in air, was soon swept away. Schrödinger himself, then Carl Eckart in the United States, and then Paul Dirac in England, proved that matrix mechanics and wave mechanics, though wildly different-looking on paper, were in fact exactly equivalent. They described the same underlying truth–whatever that might be.

Central to Schrödinger's scheme is something called the wave function, which is a mathematical object usually represented by the Greek letter psi (ψ). Over time, psi evolves in a way that's governed by Schrödinger's wave equation.

*It also appears that Davisson and a young collaborator named Charles Kunsman observed electron diffraction in 1923–four years before the famous Davisson and Germer experiment–without realizing it.

And while it evolves, it contains within it all the possibilities of the future of the quantum system that it describes, whether that system be an atom, a large or small group of subatomic particles, or a single particle. If a quantum system is "closed," as in the case of an electron confined within a box of some kind, its wave function remains constant over time, just as the standing wave of a plucked string is fixed. But if a system is "open," as when an electron moves freely through space, psi is time-dependent and tends to spread out.

Schrödinger himself believed and hoped that psi stood for some kind of real wave, like one of de Broglie's pilot waves, for example. In the case of an electron, he suggested that psi might be related to how the electron's charge became smeared out as the particle moved through space. According to this idea, the spread-out charge would have to suddenly–in a timeless instant–snap back together at a single point if any effort was made to record the electron with a measuring device. But that didn't seem possible. Any instantaneous coming together would fly in the face of Einstein's special theory of relativity, which forbids information, carried either by matter or by energy, to travel faster than the speed of light.

A way around this difficulty was suggested by Max Born, a quantum explorer of the older generation. His insight, brilliant and radical, formed the basis for all future thinking in quantum mechanics. Born argued that the wave function doesn't stand for a real wave at all, but instead a *wave of probability*. In other words, the wave function gives the odds of finding an electron, or whatever it is, at a particular place or moving in a particular way if you go looking for it.

Born interpreted the wave function not as a material thing but as a description of a system's *quantum state*. If we're talking about an electron, the quantum state includes all the positions and momenta that the electron *might* have, together with a weighting for each of these possible configurations. Another way to think of the wave function is as the collection of all alternatives available to the electron, in terms of location and movement, together with a measure of the likelihood that each of these alternatives will be played out. In a nutshell, it's a probability distribution.

Psi is the nearest quantum equivalent to what in everyday physics would be called a trajectory. But whereas a classical trajectory is a single path, unique and predetermined once the initial conditions are set, a quantum trajectory, as conceived by Born, is made up of many different ghostly alternatives, some more likely than others. To pin down the particle and find out where it actually is or how it's actually moving we have to look for it–that is, we have to carry out an observation or measurement. When we do this, Born says, the probability of the particle showing up at a particular place or with a particular momentum is given by the square of the amplitude of psi (written as $|\psi|^2$) for that particular position or momentum. Why the square of the amplitude of psi and not psi itself? Because psi is a complex function–complex in the mathematical sense that it contains terms that involve the square root of minus one, $\sqrt{-1}$ (written as i). Such functions, though meaningful mathematically, don't correspond to quantities that can be actually measured. Using the amplitude (also known as the modulus) is a process that gets rid of all the i's. Squaring the amplitude makes sure that the result is positive so that the result corresponds to a valid measure of probability.

As Born saw it, particles exist and are fully described between measurements by a quantum state represented mathematically by psi. They, or their properties, aren't actually spread out across space as Schrödinger thought: Born assumes instead that the particle really is localized somewhere. But until we seek it out, the best information available about where it is and how it's moving is the probabilistic information given by psi. At any moment in time, the probability of the particle showing up, during a measurement, at a certain place or with a certain momentum is given by the value of $|\psi|^2$ corresponding to that location or momentum. The complete set of values of $|\psi|^2$ is, so to speak, a measure of the intensity of presence of the quantum object across physical space and across a mathematical "momentum space," each point of which corresponds to a unique momentum.

Born was the first to come out and say that a fundamental randomness is built into the laws of nature. And he was also the first to point out what this inherent randomness implies: the notion of causality (that every event is the direct result of some preceding event), which is axiomatic in traditional physics, has to be abandoned in the quantum domain. In 1926, referring to collisions at the quantum level, Born wrote: "One does not get an answer to the question, What is the state after collision? but only to the question, How probable is a given effect of the collision? From the standpoint of our quantum mechanics, there is no quantity which causally fixes the effect of a collision in an individual event." These Bohemian notions of probability waves and lack of causality didn't sit well with some physicists. They didn't sit at all well with Einstein who, although he did so much to open the doors to the quantum realm, could never accept that at the heart of reality lay pure randomness and chance. "God does not play dice," he famously insisted.

But there were harder times to come for those hoping for

a swift return to the sanity of a deterministic universe. The remarkable new reading of wave mechanics by Born was quickly taken up by another senior figure, Niels Bohr (now, like Born, in his forties), and given a further twist. In 1916, Bohr had already made an important contribution (aside from his atomic model) to shaping thought about the quantum world with his "correspondence principle." This insists, reasonably enough, that quantum theory has to agree with classical theory in situations where classical theory is accurate. It also points to the inevitable fact that there must be some interface or no man's land–still to be properly explored–where the quantum and everyday regimes meet and their differences are reconciled. In 1927, Bohr added to quantum lore his "complementarity principle," which argues that the particle and wave aspects of quantum phenomena aren't contradictory. Instead, they're *complementary* descriptions of the same reality. The aspect that's revealed, particle-like or wavelike, is decided by the type of experiment we choose to carry out. No experiment can be devised, said Bohr, that will ever catch an electron, say, showing both the particle and wave sides of its nature at the same time.

Bohr embraced Born's vision of the wave function as a wave of probability. But he went much further in speculating about what this meant in terms of the nature of reality. In Born's view, a quantum object, such as an electron, is a particle even when we aren't looking; in other words, there's a definite physical entity behind the probabilistic description given by psi. Bohr disagreed. An electron might *behave* as if it were a particle or a wave as the result of a measurement, but its true nature is neither of these things nor even a combination of them. According to Bohr, unless and until we carry out an observation, an entity like an electron can't be said to exist at any place or in any material form at all. In between our attempts to pin it down, it has no existence, no substance, outside the mathematics of the wave function. It isn't, Bohr insisted, that we lack the technical means to keep track of it, instant by instant, or that our theories aren't yet precise enough to work out just where the electron is, or even that quantum mechanics throws a veil over its whereabouts. The fact is, according to Bohr, there *are* no electrons or photons or any other kinds of quantum objects out there until we go looking for them. Only in the act of measurement does the possible become real.

This extraordinary, almost surreal worldview grew out of conversations mostly between Bohr and Heisenberg in Denmark around 1926, and for this reason became known as the Copenhagen interpretation. For many years it was the standard way of thinking about how the universe works at the quantum level, and it still attracts plenty of support today. To be fair, the vast majority of scientists who use quantum mechanics for practical purposes, and even many pure quantum theorists, don't worry too much about the metaphysics of the subject or about the reality or unreality that lies behind the equations they solve. Most physicists, it's probably true to say, just get on with their physics, and save any philosophical thoughts they may have for the coffeehouse or the local bar. The Copenhagen interpretation became orthodox, not because it can be rigorously proven (it can't), but because it was established early on; it was largely the brainchild of Bohr, who was well revered (possibly more than he should have been), and few people bothered to question it. Operationally, in the vast majority of experiments, it doesn't matter whether it's right or wrong. However, at a deep level, the true nature of the subatomic world and how we interact with it matters a great deal. It also matters when we come to think about teleportation and its applications.

To get a better handle on the Copenhagen interpretation, think of the double-slit experiment that we talked about in the last chapter. And, for the sake of argument, assume we're using electrons instead of light. Following the discovery that wave-particle duality affects matter in the same way it does radiation, we know that electrons, in passing through the two slits of the double-slit experiment, will give rise to an interference pattern. In terms of wave mechanics, this is due to the *superposition*–the adding together and taking away–of the wave functions of all the individual electrons to give an overall wave function for the system.

Now focus on the single electron case. Only one electron is allowed to be in flight through the apparatus at any time. We know from our earlier discussion involving photons what will happen. The resulting interference pattern will look exactly like that seen when many electrons at once are allowed to travel from source to destination. The only difference is that in the single electron case the pattern will build up one point at a time. How can this be? And how would Bohr have interpreted the outcome?

Keep in mind that the wave function effectively encapsulates all the possible conditions in which a quantum system could *be*–the totality of what *might be* for that system. In the case of a single electron in the double-slit experiment, the wave function includes all the possibilities as to where the electron might be found (or with what momentum it might be found, if we choose to measure momentum) and the various likelihoods of it turning up in these (infinitely many) different locations if an attempt is made to look for it. Crucially, the wave function isn't just a compendium of individual destinies–a list of Newtonian trajectories–each of which is independent of all other possibilities. It's a compendium of destinies each of which takes into account, and is influenced by, every other possible outcome. This holistic, interwoven nature of the components of the wave function is the key to understanding the curious result of the double-slit experiment and the key to the strange goings-on of the quantum world itself. The wave function of the experiment doesn't include contributions from cases in which an electron travels through only one or the other of the slits. These would be storylines in isolation–classical trajectories that stood without regard to how else the electron might behave. The wave function includes only eventualities that stem from a *mutual interaction* of the alternative possible states of the electron.

Understanding the nature of the wave function to be a web of interconnected possibilities helps explain why the interference pattern emerges in the double-slit experiment whether we're dealing with one electron at a time or with many. The wave function of each electron describes the complete ensemble of possible locations and momenta, each weighted according to its probability, that is available. In other words, the probability distribution for a single electron is identical to that for a very large number of electrons. Consequently, the same pattern builds up; it just takes longer.

We can thus interpret the observed interference pattern in the double-slit experiment as a kind of printout of a cross section of the probability distribution for the position of the electron–whether we're dealing with one electron at a time or with many. But there's a deeper mystery that makes itself felt particularly in the single electron case. If an electron has no reality outside the wave function, as the Copenhagen interpretation maintains, what causes the sudden materialization of a real particle at a specific place when an observation is made? Something dramatic, it seems, happens at that moment of registration. The wave function collapses (physicists talk more prosaically about "state vector reduction"), and the many possibilities, or potentialities, of the system abruptly give way to a single actuality.

It's easy to get the impression that everything at the quantum level is fuzzy and indeterminate. But it isn't. The fact is that Schrödinger's wave equation gives a *completely deterministic* evolution of the wave function, once psi is specified at any one time. It's true that psi doesn't correspond to a unique trajectory of the type found in Newtonian physics. But there's nothing nebulous or ill-defined about it. Quite the opposite: it's a very precise mathematical description. All of the indeterminacy, all of the mystery and uncertainty, comes in at the point of observation. Only in that instant, according to the Copenhagen interpretation, does the wave function telescope down, or collapse, from its continuous and smoothly evolving form into a single outcome, effectively plucked out of the quantum hat–the outcome that we observe.

The act of measurement somehow prompts nature to make a choice, on the spur of the moment, from among all the ghostly possibilities of a quantum system–a choice made randomly, with the probabilities determined by the wave function. Suddenly, the observer sees one definite classical state, and from then on only that part of the wave function survives to continue its evolution.

In the Copenhagen view of affairs, an observation or measurement forces nature to roll its die and show us the result. Unwatched, the world keeps open all its options–all the possibilities that don't exclude other possibilities. But the act of observing a quantum system compels the random selection of a single outcome. Instead of a quantum state corresponding to a probability distribution, there's suddenly a physical entity at a definite place (or with a definite momentum). In the case of the single-electron double-slit experiment, all the ghost electrons that we might imagine are interacting to make the interference pattern coalesce into a solitary spot, a realized particle, here and now.

The Copenhagen interpretation attaches a very particular and profound meaning to the collapse of the wave function. When a spot appears on the recording surface of the single-electron double-slit experiment, we're not to think of this as calling our attention to a particle whose whereabouts were previously unknown (as Born might have argued). Instead, prior to the observation of the spot, *there was no particle.* So it isn't correct to claim that an electron collided with the detector and produced a visible trace. Bohr's version of events is that the observing apparatus interacted with the electron's wave function causing it to implode to a single point in space and time.

This interaction between measurer and measuree, however, must be of a special kind. The only way the wave function can be made to collapse, according to the Copenhagen scenario, is if the observing apparatus is considered *not to be part of the quantum state* of the system being observed. If it were otherwise, the apparatus would be incorporated within a larger wave function that embraced the quantum state of the system it was supposed to be measuring, and it's impossible for a wave function to bring about its own collapse. This impossibility was demonstrated theoretically by the brilliant Hungarian-American mathematician John von Neumann in the 1930s. The very meaning of this type of measurement that implodes wave functions refers to our drawing a distinction between the subatomic, level where quantum rules apply, and the macroscopic realm under the jurisdiction of classical physics, where irreversible changes take place and traces are recorded. Whatever does the observing must stand outside the wave function being interrogated and must, in fact, be regarded as immune to the laws of quantum mechanics.

Yet there's a complication in making this essential distinction between what is observing and what is being observed. Everything in nature ultimately consists of subatomic particles that obey quantum rules. So we're left to wonder where, along the ladder of complexity, from electron to human brain, the threshold is reached at which something becomes capable of wave function collapse. Does it have to involve a sentient observer, as von Neumann believed and Bohr suspected? Or can an inanimate piece of laboratory equipment do the job on its own? And if a sentient observer is needed, can it be an ape or a young child instead of an intelligent human adult? Would perhaps a smart computer do? Or does it take a community of savvy humans, at least some of whom understand the principles of quantum mechanics and can knowledgeably interpret the results that their instruments record?

Many different opinions have been voiced. But they are only opinions, and in the final analysis, the Copenhagen interpretation is silent on this delicate issue. Nor does the Copenhagen interpretation have anything to say about what actually causes the wave function to change so abruptly. Wave function collapse–also called state vector reduction–is taken as an article of faith, without any equation to specify how or when it happens.

Despite these failings, Bohr's conception stood as the prevailing paradigm for several decades. And although today it's only one of several schools of thought on the issue of quantum reality, it has benefited from some observational support. In 1977, for example, researchers at the University of Texas showed theoretically that the decay of an unstable particle, such as a radioactive nucleus, is suppressed by the act of observation. The more times it's observed, the greater is the suppression. And if it's observed continuously, the decay simply doesn't happen. Each observation effectively stops the spread of the wave function toward the decayed state and makes it snap back toward a sharply defined, undecayed state. This has an astonishing implication: a radioactive nucleus that is watched all the time remains intact forever, despite the fact that it's intrinsically unstable. In 1990 an experiment at the National Institute of Standards and Technology in Boulder, Colorado, was performed using several thousand beryllium ions; in this experiment, a transition from a low-energy to a high-energy state, which should have been stimulated by a radio frequency field, was prevented from happening by closely monitoring the ions with short pulses of light.[28] Apparently, the quantum equivalent of "a watched pot never boils" is more than an old physicist's tale.

Despite such evidence, the Copenhagen interpretation, with its emphasis on the role of the observer–or, at least, the observing apparatus–isn't an easy pill to swallow. Einstein flatly rejected Bohr's opinion that atoms, electrons, photons, and so forth, exist only at the instant of observation. "I cannot imagine," he said, "that a mouse could drastically change the universe merely by looking at it. . . . The belief in an external world independent of the perceiving subject is the basis of all natural science."

Schrödinger, too, was uneasy. He realized that the effects of the Copenhagen interpretation could ripple all the way up to the everyday world, with consequences that seemed to defy reason. To illustrate this point he created a thought experiment in which an unfortunate cat is put into a room in which there's a flask of poisonous gas. Poised over the flask is a hammer. By way of a detection and amplification system, the hammer will fall and break the flask, whose freed contents will kill the cat, but only if a single radioactive nucleus stored nearby decays. The behavior of the nucleus is described by a wave function that is the superposition or simultaneity of two states, decayed and not-decayed. In this unobserved condition, it's meaningless from the Copenhagen viewpoint to ask whether the decay has or hasn't happened. The only reality is the probabilistic description of the wave function, which allows for the potentiality of both realities.

That's all very well. But the fate of the nucleus decides the fate of the cat. So what we really have is a situation in which there's a superposition of quantum states covering both possibilities for the cat: a dead-cat state and a live-cat state. Only by making an observation, by looking into the room, can we force nature to make a choice. Until we do, the cat is neither alive nor dead. In fact, because we're treating it as part of a large quantum system, the Copenhagen interpretation denies it has any material reality at all until a higher system–a macroscopic measuring system–looks in upon it.

Difficulties such as Schrödinger's cat and the lack of a rationale for wave function collapse led eventually to rivals to the Copenhagen scheme. One of these, announced in 1957, was the brainchild of Hugh Everett III, then a graduate student at Princeton. Everett sidestepped the whole issue of how and why wave functions collapse by arguing that they *never* collapse. In his view, every time an observation is made, the wave function simply splits apart to take account of all the possible outcomes. No longer does the observer stand outside the system but instead enters a superposition of the various future realities. Whereas the Copenhagen interpretation leads to a breakdown of determinism because it allows us to predict only the probability of what we eventually observe, Everett's proposal goes out of its way to preserve determinism: each of the possible states, it maintains, really occurs.

This so-called many-worlds interpretation leads to the bizarre notion of the universe continually dividing into a number of alternatives, which then pursue separate evolutionary paths. Since this division must happen an enormous number of times every second and has presumably been going on for billions of years, the result is truly mind boggling. Somewhere, sometime, among all the host of continually dividing parallel universes, everything that can happen, does happen.

For some physicists, particularly those who work in the nascent field of quantum computing, Everett's multiverse model has proved surprisingly attractive. Others have balked at its spectacular lack of economy. In any event, since about 1970, scientists have been gravitating toward yet another theoretical model that seeks to save the macroscopic world from being invaded by quantum superposition. This latest scheme, called decoherence, while not completely replacing the need for an underlying interpretation such as Copenhagen or many-worlds, does help explain why the world around us looks classical rather than quantum.

Decoherence is an effect that arises from the fact that all systems interact with their surroundings: for example, by giving off and absorbing heat in the form of thermal photons.

Through these interactions, it's supposed, the multitude of quantum possibilities wash out of the microworld, giving rise to the familiar and reassuringly solid world of everyday life. Since any system must communicate with its environment, even if it does so very tenuously, the macroscopic world is shaped into distinct occurrences.

In 2004, evidence of decoherence was obtained for the first time in the laboratory by Anton Zeilinger and his colleagues in Vienna.[25] Zeilinger's team fired molecules of carbon-70 (miniature geodesic structures named fullerenes or buckyballs, after the architect Buckminster Fuller) in a vacuum, toward punctured pieces of gold foil. The slits in the foil, looking like the gaps between bars in a prison cell, were 475 nanometers (billionths of a meter) wide. Each carbon-70 molecule is about one nanometer across; however, its de Broglie wavelength–the length of the quantum wave associated with it–is 1,000 times smaller at about 3 picometers (trillionths of a meter). The gap width in the gold foil was chosen so that as each fullerene molecule reached the grating, its wavelike characteristics were engaged, causing it to diffract or to radiate out as if each aperture were the source of the wave. After 38 centimeters, this spreading wave hit a second, identical grating, passed through multiple apertures side by side, and diffracted again as it passed through each slit. On the other side, the emerging waves interfered, exactly as in the classic double-slit experiment (of which this is a pocket-sized version), producing the familiar zebra stripes of an interference pattern. Carbon-70 molecules are among the largest objects ever to have been seen showing quantum wave properties.

To start with, Zeilinger and his team heated the buckyballs to a temperature of about 1,200°C before firing them at speeds of around 190 meters per second toward the gold foils and the detector beyond. But then they cranked up the temperature and made an interesting discovery. Above 1,700°C or so, some of the fullerene molecules started smacking into the previously pure white bands of the interference pattern–their particlelike qualities becoming more asserted. As the temperature rose further, the zebra stripes grew less and less distinct until, at about 2,700°C, all that was left was a smooth elephant gray. Decoherence had become complete, destroying the interference pattern and blurring over the view of the quantum world.

This same effect also takes some of the mystery out of the Schrödinger's cat affair. Information about the state of the cat, in the form of heat energy arising from its metabolism if it's still alive, will inevitably leak out of the box allowing anyone on the outside to determine the cat's condition without actually looking in. At the quantum level, this interaction of the cat with the external environment leads to decoherence in the wave function of the cat-atom system, forcing the state out of its ghostly superposition.

Einstein once turned to his colleague Abraham Pais and asked, "Do you really believe that the moon only exists when you're looking at it?" Decoherence may help save the moon from Bohr's shadowland of the unobserved. But another quantum mystery to which Einstein vehemently objected is very much here to stay.

3

The Mysterious Link

Identical twins, it's said, can sometimes sense when one of the pair is in danger, even if they're oceans apart. Tales of telepathy abound. Scientists cast a skeptical eye over such claims, largely because it isn't clear how these weird connections could possibly work. Yet they've had to come to terms with something that's no less strange in the world of physics: an instantaneous link between particles that remains strong, secure, and undiluted no matter how far apart the particles may be–even if they're on opposite sides of the universe. It's a link that Einstein went to his grave denying, yet its existence is now beyond dispute. This quantum equivalent of telepathy is demonstrated daily in laboratories around the world. It holds the key to future hyperspeed computing and underpins the science of teleportation. Its name is *entanglement.*

• • •

The concept, but not the name, of entanglement was first put under the scientific spotlight on May 15, 1935, when a paper by Einstein and two younger associates, Nathan Rosen and Boris Podolsky, appeared in the journal *Physical Review*.[20] Its title—"Can a Quantum-Mechanical Description of Physical Reality Be Considered Complete?"—leaves no doubt that the paper was a challenge to Bohr and his vision of the subatomic world. On June 7, Schrödinger, himself no lover of quantum weirdness, wrote to Einstein, congratulating him on the paper and using in his letter the word *entanglement*—or, rather, its German equivalent *verschränkung*—for the first time. This new term soon found its way into print in an article Schrödinger sent to the Cambridge Philosophical Society on August 14 that was published a couple of months later.[43] In it he wrote:

> When two systems . . . enter into temporary physical interaction . . . and when after a time of mutual influence the systems separate again, then they can no longer be described in the same way as before, viz. by endowing each of them with a representative of its own. I would not call that one but rather the characteristic trait of quantum mechanics, the one that enforces its entire departure from classical lines of thought. By the interaction the two representatives [the quantum states] have become entangled.

The characteristic trait of quantum mechanics . . . the one that enforces its entire departure from classical lines of thought—here was an early sign of the importance attached to this remarkable effect.

Entanglement lay at the very heart of quantum reality–its most startling and defining feature. And Einstein would have none of it.

For the best part of a decade, the man who revealed the particle nature of light had been trying to undermine Bohr's interpretation of quantum theory. Einstein couldn't stomach the notion that particles didn't have properties, such as momentum and position, with real, determinable (if only we knew how), preexisting values. Yet that notion was spelled out in a relationship discovered in 1927 by Werner Heisenberg. Known as the uncertainty principle, it stems from the rule, which we met earlier, that the result of multiplying together two matrices representing certain pairs of quantum properties, such as position and momentum, depends on the order of multiplication. The same oddball math that says X times Y doesn't have to equal Y times X implies we can never know simultaneously the exact values of both position and momentum. Heisenberg proved that the uncertainty in position multiplied by the uncertainty in momentum can never be smaller than a particular number that involves Planck's constant.* In one sense, this relationship quantifies wave–particle duality. Momentum is a property that waves can have (related to their wavelength); position is a particlelike property because it refers to a localization in space. Heisenberg's formula reveals the extent to which one of these aspects fades out as the other becomes the focus of attention. In a different but related sense, the uncertainty principle tells how much the complementary descriptions of a quantum object

*Heisenberg's uncertainty relation for position and momentum is $\Delta p \cdot \Delta q \geq h/2\pi$, where Δp and Δq are the uncertainties in position and momentum (Δ is the Greek capital delta), respectively, and h is Planck's constant.

overlap. Position and momentum are complementary properties because to pin down one is to lose track of the other: they coexist but are mutually exclusive, like opposite sides of the same object. Heisenberg's formula quantifies the extent to which knowledge of one limits knowledge of the other.

Einstein didn't buy this. He believed that a particle *does* have a definite position and momentum all the time, whether we're watching it or not, despite what quantum theory says. From his point of view, the Heisenberg uncertainty principle isn't a basic rule of nature; it's just an artifact of our inadequate understanding of the subatomic realm. In the same way, he thought, wave–particle duality isn't grounded in reality but instead arises from a statistical description of how large numbers of particles behave. Given a better theory, there'd be no wave–particle duality or uncertainty principle to worry about. The problem was, as Einstein saw it, that quantum mechanics wasn't telling the whole story: it was incomplete.

Intent on exposing this fact to the world and championing a return to a more classical, pragmatic view of nature, Einstein devised several thought experiments between the late-1920s and the mid-1930s. Targeted specifically at the idea of complementarity, these experiments were designed to point out ways to simultaneously measure a particle's position *and* momentum, or its precise energy at a precise time (another complementary pair), thus pulling the rug from under the uncertainty principle and wave–particle duality.

The first of these gedanken was talked about informally in 1927, in hallway discussions at the fifth Solvay Conference in Brussels. Einstein put to Bohr a modified version of the famous double-slit experiment in which quantum objects–electrons, say–emerging from the twin slits are observed by shining light onto them. Photons bouncing off a particle would have their momenta changed by an amount that would reveal the particle's trajectory and, therefore, which slit it had passed through. The particle would then go on to strike the detector screen and contribute to the buildup of an interference pattern. Wave–particle duality would be circumvented, Einstein argued, because we would have simultaneously measured particlelike behavior (the trajectory the particle took) and wavelike behavior (the interference pattern on the screen).

But Bohr spotted something about this thought experiment that Einstein had overlooked. To be able to tell which slit a particle went though, you'd have to fix its position with an accuracy better than the distance between the slits. Bohr then applied Heisenberg's uncertainty principle, which demands that if you pin down the particle's position to such and such a precision, you have to give up a corresponding amount of knowledge of its momentum. Bohr said that this happens because the photons deliver random kicks as they bounce off the particle. The result of these kicks is to inject uncertainty into the whereabouts of the particle when it strikes the screen. And here's the rub: the uncertainty turns out to be roughly as large as the spacing between the interference bands. The pattern is smeared out and lost. And with it disappears Einstein's hoped-for contradiction.

On several other occasions, Einstein confronted Bohr with thought experiments cunningly contrived to blow duality out of the water. Each time, Bohr used the uncertainty principle to exploit a loophole and win the day against his archrival (and, incidentally, good friend). In the battle for the future of quantum physics, Bohr defeated Einstein and, in the process, showed just how important Heisenberg's little formula was in the quantum scheme of things.

Such is the version of this clash of twentieth-century titans that's been dutifully repeated in textbooks and spoon-fed to physics students for many years. But evidence has recently come to light that Bohr had unwittingly hoodwinked Einstein with arguments that were fundamentally unsound. This disclosure doesn't throw quantum mechanics back into the melting pot, but it does mean that the record needs setting straight, and that the effect that really invalidates Einstein's position should be given proper credit.

The revisionist picture of the Bohr–Einstein debates stems partly from a suggestion made in 1991 by Marlan Scully, Berthold-Georg Englert, and Herbert Walther of the Max Planck Institute for Quantum Optics in Garching, Germany.[44] These researchers proposed using *atoms* as quantum objects in a version of Young's two-slit experiment. Atoms have an important advantage over simpler particles, such as photons or electrons: they have a variety of internal states, including a ground (lowest energy) state and a series of excited (higher energy) states. These different states, the German team reckoned, could be used to track the atom's path.

Seven years later, Gerhard Rempe and his colleagues at the University of Konstanz, also in Germany, brought the experiment to life–and made a surprising discovery.[19] Their technique involved cooling atoms of rubidium down to within a hair's breadth of absolute zero. (Cold atoms have long wavelengths, which make their interference patterns easier to observe.) Then they split a beam of the atoms using thin barriers of pure laser light. When the two beams were combined, they created the familiar double-slit interference pattern. Next, Rempe and his colleagues looked to see which path the atoms followed. The atoms going down one path were left alone, but those on the other path were nudged into a higher energy state by a pulse of microwaves (short-wavelength radio waves). Following this treatment, the atoms, in their internal states, carried a record of which way they'd gone.

The crucial factor in this version of the double-slit experiment is that microwaves have hardly any momentum of their own, so they can cause virtually no change to the atom's momentum–nowhere near enough to smear out the interference pattern. Heisenberg's uncertainty principle can't possibly play a significant hand in the outcome. Yet with the microwaves turned on so that we can tell which way the atoms went, *the interference pattern suddenly vanishes.* Bohr had argued that when such a pattern is lost, it happens because a measuring device gives random kicks to the particles. But there aren't any random kicks to speak of in the rubidium atom experiment; at most, the microwaves deliver momentum taps ten thousand times too small to destroy the interference bands. Yet, destroyed the bands are. It isn't that the uncertainty principle is proved wrong, but there's no way it can account for the results.

The only reason momentum kicks seemed to explain the classic double-slit experiment discussed by Bohr and Einstein turns out to be a fortunate conspiracy of numbers. There's a mechanism at work far deeper than random jolts and uncertainty. What destroys the interference pattern is the very act of trying to get information about which path is followed. The effect at work is entanglement.

Ordinarily, we think of separate objects as being independent of one another. They live on their own terms, and anything tying them together has to be forged by some tangible physical mechanism. But it's a different story in the quantum world. If a particle interacts with some object–another particle, perhaps–then the two can become inextricably joined, or entangled. In a sense, they stop being independent things and can only be described in relation to each other, as if joined by a psychic link.

This has profound consequences for a particle's ability to show wavelike behavior. By itself, an atom can act as a wave. In a two-slit device, however, it effectively splits its own existence and goes through both slits. If these two phantoms of the atom move along their paths without running into anything, then they recombine and interfere at the screen.

But suppose you send a photon toward one of the slits to check out what's happening. If an atom were there, the photon would simply careen off and record the atom's position. But because the atom's identity is already split between the two paths, it makes the photon split too. A ghost of the photon bounces off the ghost of the atom at that slit, and a second photon ghost carries straight on. Entanglement is now in business: the interaction pairs up each ghost atom with a corresponding ghost photon. Tied to their photon parasites, the two atom ghosts are no longer matched, so the interference vanishes.

For several decades, physicists labored under the delusion that what makes quantum theory so weird is its inherent uncertainty or fuzziness. But all along it was really entanglement that lay at the root of the mystery. Loss of interference, as happens when we ask which path a particle follows through the double-slit experiment, is *always* due to entanglement. Quantum particles can split into ghosts that can move on many paths at once, and when they come back together we see wavelike behavior and interference patterns. But reach into the quantum world, and you can't avoid attaching disruptive partners to the quantum ghosts–partners that will spoil the reunion and make the ghosts act as if they were true particles.

Bohr defeated Einstein, and rightly so. But he defeated him for the wrong reason. In drawing so heavily on the uncertainty principle, he persuaded the scientific community that Heisenberg's relationship was more important than it really is. The irony is that, even more than with quantum uncertainty and indeterminism, Einstein wouldn't countenance the effect that tied quantum particles together by what he called "spooky action at a distance." His famous attack on the problem, through what became known as the Einstein-Podolsky-Rosen (EPR) paradox, sent tremors across the Atlantic and down the corridors of time to the present day.

In 1935, having immigrated to the United States two years earlier, Einstein was settled into his position as professor of theoretical physics at the newly created Institute for Advanced Studies in Princeton. He'd been quiet on the quantum front for a while: his last gedanken challenge to Bohr had been in 1930. But now, with the help of two research assistants, the Russian Boris Podolsky and the American-born Israeli Nathan Rosen, he was ready to launch his greatest and final assault on the foundations of the quantum realm.

Einstein and his colleagues dreamed up a situation that led to such a strange–and, in their opinion, completely unacceptable–conclusion that it proved for them that there was something wrong or missing from the quantum mechanical view of the world. In a four-page paper, they envisioned two particles, *A* and *B*, which have come into contact, interacted for a brief while, and then flown apart. Each particle is described by (among other properties) its own position and momentum. The uncertainty principle insists that one of these can't be measured precisely without destroying knowledge of the other. However, because *A* and *B* have interacted and, in the eyes of quantum physics, have effectively merged to become one interconnected system, it turns out that the momentum of both particles taken together and the distance between them can be measured as precisely as we like. Suppose we measure the momentum of *A*, which we'll assume has remained behind in the lab where we can keep an eye on it. We can then immediately deduce the momentum of *B* without having to do any measurement on it at all. Alternatively, if we choose to observe the position of *A*, we would know, again without having to measure it, the position of *B*. This is true whether *B* is in the same room, on Mars, or far across the reaches of intergalactic space.

From Heisenberg's relationship, we know that measuring the position of, say, *A* will lead to an uncertainty in its momentum. Einstein, Podolsky, and Rosen pointed out, however, that by measuring the position of *A*, we gain precise knowledge of the position of *B*. Therefore, if we take quantum mechanics at face value, by gaining precise knowledge of its position, an uncertainty in momentum has been introduced for *B*. In other words, the state of *B* depends on what we *choose* to do with *A* in our lab. And, again, this is true whatever the separation distance may be. EPR considered such a result patently absurd. How could *B* possibly know whether it should have a precisely defined position or momentum? The fact that quantum mechanics led to such an unreasonable conclusion, they argued, showed that it was flawed–or, at best, that it was only a halfway house toward some more complete theory.

At the core of EPR's challenge is the notion of *locality*: the commonsense idea that things can only be affected directly if they're nearby. To change something that's far away, there's a simple choice: you can either go there yourself or send some kind of signal. Either way, information or energy has to pass through the intervening space to the remote site in order to affect it. The fastest this can happen, according to Einstein's special theory of relativity, is the speed of light.

The trouble with entanglement is that it seems to ride roughshod over this important principle. It's fundamentally *nonlocal*. A measurement of particle *A* affects its entangled partner *B* instantaneously, whatever the separation distance, and without any signal or influence passing between the two locations. This bizarre quantum connection isn't mediated by fields of force, like gravity or electromagnetism. It doesn't weaken as the particles move apart, because it doesn't actually stretch across space. As far as entanglement is concerned, it's as if the particles were right next to one another: the effect is as potent at a million light-years as it is at a millimeter. And because the link operates outside space, it also operates outside time. What happens at *A* is immediately known at *B*. No wonder Einstein used words such as "spooky" and "telepathic" to describe–and deride–it. No wonder that as the author of relativity he argued that the tie that binds entangled particles is a physical absurdity. Any claim that an effect could work at faster-than-light speeds, that it could somehow serve to connect otherwise causally isolated objects, was to Einstein an intellectual outrage.

A close look at the EPR scenario reveals that it doesn't actually violate causality, because no information passes between the entangled particles. The information is already, as it were, built into the combined system, and no measurement can add to it. But entanglement certainly does throw locality out the window, and that development is powerfully counterintuitive. It was far too much for Einstein and his two colleagues to accept, and they were firmly convinced that such ghostly long-range behavior meant that quantum mechanics, as it stood, couldn't be the final word. It was, they suggested, a mere approximation of some as yet undiscovered description of nature. This description would involve variables that contain missing information about a system that quantum mechanics doesn't reveal, and that tell particles how to behave before a measurement is carried out. A theory along these lines–a theory of so-called local hidden variables–would restore determinism and mark a return to the principle of locality.

The shock waves from the EPR paper quickly reached the shores of Europe. In Copenhagen, Bohr was once again cast into a fever of excitement and concern as he always was by Einstein's attacks on his beloved quantum worldview. He suspended all other work in order to prepare a counterstrike. Three months later, Bohr's rebuttal was published in the same American journal that had run the EPR paper. Basically, it argued that the nonlocality objection to the standard interpretation of quantum theory didn't represent a practical challenge. It wasn't yet possible to test it, and so physicists should just get on with using the mathematics of the subject, which worked so well, and not fret about the more obscure implications.

Most scientists, whose interest was simply in using quantum tools to probe the structure of atoms and molecules, were happy to follow Bohr's advice. But a few theorists continued to dig away at the philosophical roots. In 1952, David Bohm, an American at Birkbeck College, London, who'd been hounded out of his homeland during the McCarthy inquisitions, came up with a variation on the EPR experiment that paved the way for further progress in the matter.[11] Instead of using two properties, position and momentum, as in the original version, Bohm focused on just one: the property known as *spin*.

The spin of subatomic particles, such as electrons, is analogous to spin in the everyday world but with a few important differences. Crudely speaking, an electron can be thought of as spinning around the way a basketball does on top of an athlete's finger. But whereas spinning basketballs eventually slow down, all electrons in the universe, whatever their circumstances, spin all the time and at exactly the same rate. What's more, they can only spin in one of two directions, clockwise or counterclockwise, referred to as *spin-up* and *spin-down*.

Bohm's revised EPR thought experiment starts with the creation, in a single event, of two particles with opposite spin. This means that if we measure particle *A* and find that it's spin-up, then, from that point on, *B* must be spin-down. The only other possible result is that *A* is measured to be spin-up, which forces *B* to be spin-down. Taking this second case as an example, we're not to infer, says quantum mechanics, that *A* was spin-up *before* we measured it and therefore that *B* was spin-down, in a manner similar to a coin being heads or tails. Quantum interactions always produce superpositions. The state of each particle in Bohm's revised EPR scenario is a mixed superposition that we can write as: psi = (*A* spin-up and *B* spin-down) + (*A* spin-down and *B* spin-up). A measurement to determine *A*'s spin causes this wave function to collapse and a random choice to be made of spin-up or spin-down. At that very same moment, *B* also ceases to be in a superposition of states and assumes the opposite spin.

This is the standard quantum mechanical view of the situation and it leads to the same kind of weird conclusion that troubled Einstein and friends. No matter how widely separated the spinning pair of particles may be, measuring the spin of one causes the wave function of the combined system to collapse instantaneously so that the unmeasured twin assumes a definite (opposite) spin state, too. The mixed superposition of states, which is the hallmark of entanglement, ensures nonlocality. Set against this is the Einsteinian view that "spooky action at a distance" stems not from limitations about what the universe is able to tell us but instead from limitations in our current knowledge of science. At a deeper, more basic level than that of wave functions and complementary properties, are hidden variables that will restore determinism and locality to physics.

Bohm's new version of the EPR paradox didn't in itself offer a way to test these radically different worldviews, but it set the scene for another conceptual breakthrough that *did* eventually lead to a practical experiment. This breakthrough came in 1964 from a talented Irish physicist, John Bell, who worked at CERN, the European center for high-energy particle research in Switzerland. Colleagues considered Bell to be the only physicist of his generation to rank with the pioneers of quantum mechanics, such as Niels Bohr and Max Born, in the depth of his philosophical understanding of the implications of the theory. What Bell found is that it makes an *experimentally observable difference* whether the particles described in the EPR experiment have definite properties before measurement, or whether they're entangled in a ghostlike hybrid reality that transcends normal ideas of space and time.

Bell's test hinges on the fact that a particle's spin can be measured independently in three directions, conventionally called *x*, *y*, and *z*, at right angles to one another. If you measure the spin of particle *A* along the *x* direction, for example, this measurement also affects the spin of entangled particle *B* in the *x* direction, but not in the *y* and *z* directions. In the same way, you can measure the spin of *B* in, say, the *y* direction without affecting *A*'s spin along *x* or *z*. Because of these independent readings, it's possible to build up a picture of the complementary spin states of both particles. Being a statistical effect, lots of measurements are needed in order to reach a definite conclusion. What Bell showed is that measurements of the spin states in the *x*, *y*, and *z* directions on large numbers of real particles could in principle distinguish between the local hidden variable hypothesis championed by the Einstein–Bohm camp and the standard nonlocal interpretation of quantum mechanics.

If Einstein was right and particles really did always have a predetermined spin, then, said Bell, a Bohm-type EPR experiment ought to produce a certain result. If the experiment were carried out on many pairs of particles, the number of pairs of particles in which both are measured to be spin-up, in both the *x* and *y* directions ("*xy* up"), is always less than the combined total of measurements showing *xz* up and *yz* up. This statement became known as Bell's inequality. Standard quantum theory, on the other hand, in which entanglement and nonlocality are facts of life, would be upheld if the inequality worked the other way around. The decisive factor is the degree of correlation between the particles, which is significantly higher if quantum mechanics rules.

This was big news. Bell's inequality, although the subject of a modest little paper and hardly a popular rival to the first Beatles tour of America going on at the same time, provided a way to tell by actual experiment which of two major, opposing visions of subatomic reality was closer to the truth.[5] Bell made no bones about what his analysis revealed: Einstein's ideas about locality and determinism were incompatible with the predictions of orthodox quantum mechanics. Bell's paper offered a clear alternative that lay between the EPR/Bohmian local hidden variables viewpoint and Bohrian, nonlocal weirdness. The way that Bell's inequality was set up, its violation would mean that the universe was inherently nonlocal, allowing particles to form and maintain mysterious connections with each other no matter how far apart they were. All that was needed now was for someone to come along and set up an experiment to see for whom Bell's inequality tolled.

But that was easier said than done. Creating, maintaining, and measuring individual entangled particles is a delicate craft, and any imperfection in the laboratory setup masks the subtle statistical correlations being sought. Several attempts were made in the 1970s to measure Bell's inequality but none was completely successful. Then a young French graduate student, Alain Aspect, at the Institute of Optics in Orsay, took up the challenge for his doctoral research.

Aspect's experiment used particles of light–photons–rather than material particles such as electrons or protons. Then, as now, photons are by far the easiest quantum objects from which to produce entangled pairs. There is, however, a minor complication concerning the property that is actually recorded in a photon-measuring experiment such as

Aspect's or those of other researchers we'll be talking about later. Both Bell and Bohm presented their theoretical arguments in terms of the particle spin. Photons do have a spin (they're technically known as spin-1 particles), but because they travel at the speed of light, their spin axes always lie exactly along their direction of motion, like that of a spinning bullet shot from a rifle barrel. You can imagine photons to be right-handed or left-handed depending on which way they rotate as you look along their path of approach. What's actually measured in the lab isn't spin, however, but the very closely related property of polarization.

Effectively, polarization is the wavelike property of light that corresponds to the particlelike property of spin. Think of polarization in terms of Maxwell's equations, which tell us that the electric and magnetic fields of a light wave oscillate at right angles to each other and also to the direction in which the light is traveling. The polarization of a photon is the direction of the oscillation of its *electric* field: up and down, side to side, or any orientation in between. Ordinarily, light consists of photons polarized every which way. But if light is passed through a polarizing filter, like that used in Polaroid sunglasses, only photons with a particular polarization–the one that matches the slant of the filter–can get through. (The same happens if two people make waves by flicking one end of a rope held between them. If they do this through a gap between iron railings only waves that vibrate in the direction of the railings can slip through to the other side.)

Aspect designed his experiment to examine correlations in the polarization of photons produced by calcium atoms–a technique that had already been used by other researchers. He shone laser light onto the calcium atoms, which caused the electrons to jump from the ground state to a higher energy level. As the electrons tumbled back down to the ground state, they cascaded through two different energy states, like a two-step waterfall, emitting a pair of entangled photons–one photon per step–in the process.

The photons passed through a slit, known as a collimator, designed to reduce and guide the light beam. Then they fell onto an automatic switching device that randomly sent them in one of two directions before arriving, in each case, at a polarization analyzer–a device that recorded their polarization state.

An important consideration in Aspect's setup was the possibility, however small, that information might leak from one photon to its partner. It was important to rule out a scenario in which a photon arrived at a polarization analyzer, found that polarization was being measured along say the vertical direction, and then somehow communicated this information to the other photon. (How this might happen doesn't matter: the important thing was to exclude it as an option.) By carefully setting up the distances through which the photons traveled and randomly assigning the direction in which the polarization would be measured while the photons were in flight, Aspect ensured that under no circumstances could such a communicating signal be sent between photons. The switches operated within 10 nanoseconds, while the photons took 20 nanoseconds to travel the 6.6 meters to the analyzers. Any signal crossing from one analyzer to the other at the speed of light would have taken 40 nanoseconds to complete the journey–much too long to have any effect on the measurement.

In a series of these experiments in the 1980s, Aspect's team showed what most quantum theorists expected all along: Bell's inequality was violated.[1] The results agreed completely with the predictions of standard quantum mechanics and discredited any theories based on local hidden variables. More recent work has backed up this conclusion. What's more, these newer experiments have included additional refinements designed to plug any remaining loopholes in the test. For example, special crystals have enabled experimenters to produce entangled photons that are indistinguishable, because each member of the pair has the same wavelength. Such improvements have allowed more accurate measurements of the correlation between the photons. In all cases, however, the outcomes have upheld Aspect's original discovery. Entanglement and nonlocality are indisputable facts of the world in which we live.

One of the most astonishing conclusions that follows from this finding is that the entire universe is intimately connected at the subatomic level. Some 14 billion years ago, all the matter now scattered across the vast reaches of space was huddled together inside a ball far smaller than the period at the end of this sentence. The mutual set of interactions between particles–the overlapping of their individual wave functions–at this early stage ensured that entanglement was, and is, a cosmoswide phenomenon. Even today, with matter flung across billions of light-years, there remains in place this extraordinary web linking every particle in existence. It means that if you make the slightest change to the smallest thing in existence here and now, it will have some effect, however tiny, instantaneously, throughout all known physical reality.

We can't yet tap the potential of this remarkable cosmic web, and perhaps we never will be able to for any practical purpose. But on a smaller scale, the dramatic possibilities of entanglement have already begun to be exploited. In the late

1980s, theoreticians started to see entanglement not just as a puzzle and a way to penetrate more deeply into the mysteries of the quantum world, but also as a resource. Entanglement could be exploited to yield new forms of communication and computing. It was a vital missing link between quantum mechanics and another field of explosive growth: information theory. The proof of nonlocality and the quickly evolving ability to work with entangled particles in the laboratory were important factors in the birth of a new science. Out of the union of quantum mechanics and information theory sprang quantum information science–the fast-developing field whose most stunning aspect is teleportation.

4

Dataverse

Dig down to the bedrock of reality and you get a surprise. The fundamental essence of the universe lies beyond atoms and electrons, photons and quarks. There's a new theory in town that makes an audacious claim: the basic nature of everything around us isn't matter or substance of any kind–it's *information.* One of the pioneers of this radical viewpoint is the veteran American physicist John Wheeler, a onetime colleague of Bohr and Einstein. "[T]he physical world," says Wheeler, "has at bottom–at a very deep bottom, in most instances–an immaterial source and explanation." He sums it up with the snappy little aphorism: "It comes from Bit."

We live in and are part of a cosmic tapestry woven from threads and beads–connections and nodes–of information.

That concept has emerged from a fusion of quantum mechanics and another science that's grown up alongside it, information theory. Information, it's become clear, is physical. And any processing of information is always done, and can only be done, by physical means. These reasonable and innocent-sounding conclusions have some extraordinary consequences, because at the subatomic level physical also implies quantum mechanical. So any information process is ultimately a quantum process. Therefore, if we can gain fine control over a system at the quantum level, especially if we can learn to harness the power of entanglement, we'll have at our command an astonishing tool for handling, sending, and manipulating information. That's the ambitious goal of the new discipline of quantum information science, of which teleportation and quantum computation are two of the most spectacular outgrowths.

Before quantum information science came on the scene, there was classical information theory. But even this ancestral field isn't old. Only in the first quarter of the twentieth century, with the boom in telegraphy, telephony, and other forms of electrical communication and storage, did people start catching on to the idea of information as a resource. How to convey most efficiently ever-mounting volumes of facts, figures, and spoken words; how best to record them? With the world threatened by an avalanche of data, these were becoming real and pressing issues.

The concept of information as a subject that could be tackled scientifically was first taken up in the 1920s by two electrical and communications engineers, Harry Nyquist and Ralph Hartley. Swedish-born Nyquist, who came to the

United States with his family as an eighteen-year-old in 1907, eventually earned a Ph.D. in physics from Yale, then joined American Telephone and Telegraph Company (AT&T), where he worked from 1917 to 1934, before moving to Bell Telephone Laboratories. His best-known work was done in the decade following World War I, when he was inspired by the burgeoning telegraph communications challenges of that era. Nyquist explored the theoretical limits affecting how much information could be sent by telegraph, and the frequency band, or bandwidth, needed to transmit a given amount of information. Thanks to the elegance and generality of his writings, much of it continues to be cited and used even today.

Ralph Hartley, born in Spruce, Nevada, just a few months before Nyquist, took a bachelor's degree at the University of Utah and then spent a year at Oxford as a Rhodes scholar. After this, he joined the Western Electric Company, and later Bell Labs. As early as 1915, he invented a special circuit that became known as the Hartley oscillator. But, like Nyquist, his most significant period of research came in the 1920s when he focused on the technology of electrical repeaters and, especially, voice and carrier transmission.

In a 1928 issue of the Bell Labs technical journal, both Nyquist and Hartley published papers that became classics in their field.[27, 36] Nyquist's paper "Certain Topics in Telegraph Transmission Theory" refined some earlier results he'd obtained and established the principles of sampling continuous signals to convert them to digital signals. What became known as the Nyquist sampling theorem showed that the sampling rate must be at least twice the highest frequency present in the sample in order to reconstruct the original signal.

Hartley, who'd been working completely independently of Nyquist, broke new ground in his 1928 paper by talking about "a measure of information." He developed a concept of information based on "physical as contrasted with psychological considerations." In other words, he realized that information content could be quantified in an objective way.

Hartley started from an interesting premise. He defined the information content of a message in terms of the number of messages that *might* have been sent. Take, for example, a couple of situations: the outcome of the roll of a die, which could lead to six different messages describing the result, and the outcome of drawing a card from a standard deck, in which fifty-two different messages are possible. Hartley's argument was that a message telling which side of the die came up carries less information than a message about which card was chosen at random from a deck, because in the case of the die there are fewer possibilities and less uncertainty. Hartley then gave this idea a mathematical backbone. He suggested that the way the amount of information varied with the number of possibilities available most likely depended on the logarithm* of that number. He was already familiar with logarithms cropping up all over the place in communications. "Time, bandwidth, etc.," said Hartley, "tend to vary linearly [in a straight-line way] with the logarithm of the number of possibilities: adding one relay to a group doubles the number of possible states of the

*The logarithm of a number x is the power to which another number, called the base, has to be raised to equal x. For example, the logarithm to base 10 of 100 (written as $\log_{10}100$) is 2, since $10^2 = 100$. In the same way, the logarithm to base 2 of 8 ($\log_2 8$) is 3, since $2^3 = 8$.

relays. In this sense the logarithm also feels more intuitive as a proper measure: two identical channels should have twice the capacity for transmitting information than one."

Twenty years later, another Bell Labs engineer, the wonderfully creative and gifted Claude Shannon, took the work of Nyquist and Hartley further and used it as a launching pad for a complete theory of information and communications. As a boy, Shannon had built a telegraph from his friend's house to his own out of fencing wire. Later as an adult, when his accomplishments had brought him almost legendary status, he'd still be seen riding the hallways of distinguished establishments on a unicycle, sometimes juggling–his greatest recreational passion–at the same time, or hopping along on a pogo stick. He even designed a unicycle with an off-center wheel so that it went up and down as he rode along to make juggling easier. In his "toy room" was a machine he built that had soft beanbag hands that juggled steel balls and, his ultimate masterpiece, a tiny stage on which three clowns, driven by an invisible clockwork mechanism, juggled eleven rings, seven balls, and five clubs.

Always a gadget lover and something of a loner throughout his working life, Shannon devised other wonders including a robotic mouse with copper whiskers that solved mazes in search of a piece of brass "cheese" and was one of the earliest attempts to teach a machine to learn, and a computer called the THROBAC ("THrifty ROman-numeral BAckward-looking Computer") that worked in roman numerals. Perhaps fortunately for the health of those around him, his design for a rocket-powered frisbee never left the drawing board. In 1950 he wrote a technical paper and also a popular article that appeared in *Scientific American* on the principles of programming computers to play chess. Thirty years later, he saw Bell Lab's chess-playing computer, Belle, win the International Computer Chess Championship.

Shannon graduated from the University of Michigan with a degree in electrical engineering and mathematics in 1936. Although never considered an outstanding mathematician (it was later said that he had to invent any math he needed), he'd become fascinated as a student by the work of the nineteenth-century philosopher George Boole, who devised a way of manipulating statements in logic that came to be known as Boolean algebra. Shannon transferred to MIT to pursue his master's degree and was lucky enough to find himself supervised there by Vannevar Bush, the inventor of a device called the differential analyzer, which was at the heart of all analogue computers at the time. Bush encouraged Shannon to look at ways of applying Boolean algebra to the increasingly labyrinthine problem of telephone networks. Shannon's master's thesis on the subject, written in 1936 and titled "A Symbolic Analysis of Relay and Switching Circuits," has been hailed as the most important such thesis of the twentieth century. In it the twenty-two-year-old showed how Boolean algebra could be put into action using electrical circuits of relays and switches, and how binary arithmetic, involving just zeros and ones, was the natural language of such circuits because each switch, being a two-state device that was either open or closed, could stand for a single binary digit. Few master's theses are even taken off the library shelf except perhaps by a small coterie of peers who may be researching in the same narrow field. But Shannon's effort was immediately recognized as the bridge that would lead from analogue computing to digital computing–the basis of virtually all information processing today.

In 1941, having earned a Ph.D. in mathematics at MIT and having spent a subsequent year at the Institute for Advanced Study in Princeton (where Einstein was in residence), Shannon joined Bell Labs, where he stayed for fifteen years before returning to MIT. At Bell he worked on wartime projects including cryptography (the subject of the next chapter), and developed the encryption system used by Roosevelt and Churchill for transoceanic conferences. But, unbeknownst to most of those around him, he was also laying the foundations for a broad-based theory of information and communications–work he'd begun during his year at Princeton. In 1948 the results of this momentous research emerged in a celebrated paper published in two parts in Bell Laboratory's research journal.[45] In the first paragraph of his classic discourse, Shannon acknowledges the seminal role played by the two papers written contemporaneously by Nyquist and Hartley two decades earlier.

We use the term *information* in the everyday sense to imply facts, knowledge, and meaning. It has to do with what we learn and understand from a message. But information in Shannon's information theory doesn't carry this kind of subjective, value-added connotation. Shannon wasn't interested in what a message might actually mean to someone, or even whether it made any sense at all, any more than a truck driver is interested in the contents of the load he's hauling. His big concern was with the issues involved in transmitting the message successfully and efficiently from one place to another over what might be a noisy line.

For the purposes of Shannon's theory, a message is just a string of characters–letters or numbers or both–or, in its simplest form, a string of binary digits. His paper, in fact, was one of the first to use the word *bit*, short for "binary digit," a term coined in 1946 by the Princeton mathematician John Tukey (who also gave us *byte* and *software*). As for the information content of a message, Shannon took this to be a measure of how much freedom of choice there was in selecting the message. That seems at first an odd way to think of it, tantamount to saying that information is what you *don't* know rather than what you do. But remember, we're talking here about information in the special context of transmitting a message–whatever it might consist of–from a sender to a receiver over some kind of channel, such as a telephone wire. The receiver doesn't know ahead of time what the message will be, so there's some latitude of expectation. He has the task of extracting a message–hopefully an accurate version of the original–from whatever comes down the line. And the transmission won't be perfect: fragments of the original message string are bound to get lost or garbled, and spurious stuff will be mixed in with it along the way, just as happens when two people are trying to carry on a conversation by the side of a busy road. So in terms of transmitting a message from one place to another, the key aspect of information, however strange it may seem, is uncertainty. That's why information theory seems obsessed with the messages you *might* get, or the number of messages you *could* choose from. To generalize even more, information theory deals with the statistical properties of a message, averaged out over the whole message, without any regard to its factual content or how it will be interpreted.

In information theory, the more bits of information that have to be sent, the more room for uncertainty there is in what's received. The number of possible messages you can make doubles for every extra binary digit the message contains, since each of these digits can take on one of two values: one or zero.* This means that a message with x number of binary digits could come in two to the power of x number of different forms. Flip this idea around mathematically, and the number of bits needed to transmit a message comes out to be the logarithm to base two of the number of possible messages.

This formula has a familiar ring to it, especially if you happen to be a physicist. It looks uncannily like the equation that Max Planck came up with in 1900–the so-called Boltzmann equation (because it was based on Boltzmann's ideas)–linking thermodynamic entropy to molecular disorder. In Shannon's formula, the number of possible messages takes the place of molecular disorder (which, in turn, is related to the number of possible indistinguishable arrangements of molecules), while the number of bits needed to carry the message stands in for thermodynamic entropy. Shannon had evidently exposed a deep-seated relationship between physics and information. Now it was a question, for him and others, of figuring out the implications of this surprising and fundamental link.

When entropy shows up in thermodynamics, it tells of the inexorable trend of the universe to wind down. It speaks of the inevitable descent toward cosmic anarchy and decay–inevitable because the odds always favor the unruly. Disorder

*With a single binary digit, two different messages can be sent, represented by 0 or 1. With two bits, four messages can be sent, represented by 00, 01, 10, and 11. Three bits, which can be arranged in $2^3 = 8$ possible ways, can be used to encode eight different messages, and so on.

is pandemic. People grow old and wrinkly, cars rust, desktops that aren't tidied get messier, buildings crumble. Only one thing seems to rise while all else around it declines and falls: the strange, pervasive property of entropy. A simple example reveals why.

Open a new deck of fifty-two playing cards that's ordered alphabetically by suit (clubs, diamonds, hearts, spades) then by value (two, three, . . . king, ace); toss the cards into the air and let them fall to the ground; swish them around for half a minute; then gather them up at random. Common sense tells you that it's astronomically unlikely the cards will end up in the order they started. You could throw the cards in the air all day long and patiently keep collecting them together without having the slightest hope that you'd see anything like the original order again.

In the card-tossing example, think of each arrangement of cards as being a possible *state* of the deck. Only one state matches the pristine order that the deck was in when you first took it out of its wrapper. Set against this is the huge number of other states that exist in which the cards are more or less jumbled up. The state of lowest entropy is the one with all the cards in the original order; a slightly higher-entropy state is one that has just a single card out of place. To figure out how many possible states there are altogether, remember that the first card you pick up can be any of fifty-two, the second can be any of the remaining fifty-one, the third can be any of the remaining fifty, and so on. Therefore, in total there are $52 \times 51 \times 50 \times \ldots \times 3 \times 2 \times 1$ arrangements, or, in mathematical shorthand, 52! (where the exclamation point is to be read as "factorial"). Written in full, this number comes out to be roughly 81 million trillion trillion trillion trillion trillion. Since only one of these states is in perfect order (minimum entropy) and only a few are in nearly perfect order, it's pretty clear that if the deck starts out in a state of low entropy, any shuffling is almost bound to make the entropy higher. This is exactly the same conclusion that Rudolf Clausius reached in his thermodynamic study of steam engines: in a closed system, entropy never goes down.

If fifty-two cards can be arranged in such a colossal number of ways, then imagine how much more staggering is the number of possible arrangements of a thousand different cards, or a million. Compared with the few very organized or reasonably well-organized (low-entropy) states in which the cards are close to being in suit and numerical order, the number of disorganized or nearly random states in a large deck is absolutely mind-boggling.

Now let's push this argument further to the kind of situation found in many natural physical systems. Think, for instance, about all the gas molecules in the air that fills the room you're sitting in. These molecules are dashing around in different directions at various, mostly high speeds (oxygen molecules at normal temperature average about 480 meters per second), ricocheting off one another and other objects. While in theory all the molecules could suddenly rush to one corner of the room like a cartoon cloud, leaving you to expire in astonishment in a vacuum, there's no need to panic. This stupendously well-ordered state is far less likely than the many more arrangements in which the atoms are more or less evenly spread throughout the available space. To get some idea how improbable it would be for all the air to spontaneously huddle in a little region that had, say, just 1 percent of the total volume of the room, start with a number that's twenty-seven digits long–in other words, in the hundreds of trillions of trillions. That's a ballpark figure for how many molecules of air there are in an ordinary living room. To work out the chances that all these molecules would end up in the same 1 percent of the room at the same instant, you need to multiply one hundredth by one hundredth by one hundredth and so on, for all the molecules, resulting in a number that couldn't be written out in full even if you wrote small and used all the world's stock of paper.

Max Planck was the first to quantify this type of argument, wherein the probability of a given microscopic state–the chances of a certain arrangement of molecules or atoms in a system–gives a measure of entropy. His formula (Boltzmann's equation) spells out how entropy is tied to the level of disorganization at the submicroscopic level. A highly chaotic state, which is far more likely to occur than an orderly one, corresponds to a large value of entropy.

To see how this idea carries over to information, suppose you have a jar that's full of black and white marbles. A low-entropy arrangement might consist of all the black marbles in the bottom half of the jar and all the white ones in the top half. This state can easily be distinguished from another orderly state, say one in which the black and white marbles are put in alternating layers, or one that's the reverse of the original state, with white marbles in the bottom of the jar and black marbles on top. Any of these low-entropy states, in turn, is recognizable at a glance as being distinct from a high-entropy state where the marbles have been pretty randomly distributed by shaking up the jar's contents. By contrast, assuming that, aside from the two different colors, all the marbles are essentially identical, telling one high-entropy state from another isn't nearly as easy. In fact, it can be impossible. For example, while there might be one state of the marbles with black marble *A* in the top half of the jar and black marble *B* in the lower half, all other marbles being in the same position, these two states look exactly the same. With a uniform distribution of marbles obtained by thoroughly shaking the jar, a huge number of states exist that can't be told apart. In other words, compared with low-entropy states, high-entropy states contain very little information.

This kind of logic makes perfect sense in the world of communications. A purposeful signal, one that contains a lot of useful information, is a highly ordered state. Suppose you've just e-mailed a picture of yourself to your friend in Florida. You can imagine some of the bits in the signal as coding for your hair color, others for the shading of your eyes and skin. This highly patterned collection of bits–a state containing a large amount of information–is a state of low entropy. If you took the bit content of the picture and randomly shuffled the ones and zeroes, the result isn't likely to be a new image that looks like anything familiar. Almost certainly, you'd create a complete jumble that doesn't resemble anything at all. Disorder wins again. While there are very few ways to arrange the bits into a recognizable picture of yourself (or of anything else for that matter), there are many ways of randomly picking ones and zeroes.

In comparing the mathematics and behavior of information with the mathematics and behavior of a thermodynamic system, Shannon was struck by the profound connection that existed between the two. The common factor, he realized, was entropy. That's what his formula linking the number of bits in a message to the number of possible messages was telling him. It wasn't a coincidence that this equation was a dead ringer for the one derived by Planck (Boltzmann's equation), which shows how the microscopic world of atoms and molecules is tied to the macroscopic, sensory world of heat. Entropy emerged as a crucial factor in the world of information, just as it did in the world of physics. And, most significant of all, when you looked at it closely, it played essentially the same role in both worlds. Yes, entropy was a measure of the disorderliness of a system. But above and beyond that, Shannon now appreciated, it was a measure of the extent to which a system *lacked information.* Entropy is really the negative of information so that, for example, by raising the entropy of a system–increasing its level of disorder–there's a drop-off in the amount of information that the system contains.

As of today, all this has become very well established and uncontroversial. But half a century ago, the mention of entropy in the same breath as communications, telephone lines, and bandwidths seemed very revolutionary, even bizarre. It was the great mathematician John von Neumann who finally persuaded Shannon that he really should go with the term *entropy* in this new field of information theory that he was pioneering because, as von Neumann put it, with tongue in cheek, "no one knows what entropy is, so in a debate you will always have the advantage."

Of course, von Neumann probably understood entropy as well as anyone and he, like Shannon, immediately recognized that it was a universal. In thermodynamics, it calibrates the extent to which a system is disorganized at the very small scale. Shannon's informational equivalent of entropy, or "Shannon entropy" as it's become known, reduced to its most basic terms, is the minimum number of binary digits needed to encode a message. In our twenty-first century world, awash with computers that crunch away endlessly in binary code, that sounds like a simple, even obvious way to define the information content of a message. But in 1948, at the dawn of the information age, the digitizing of information of any sort, together with the introduction of entropy, was a radical step.

Typically, Shannon entropy is measured in bits per symbol or some similar unit. Computers, for example, use what's known as the extended ASCII character set to represent letters, numbers, punctuation marks, and other common symbols. This has 256 characters, because 8 bits are assigned to represent each symbol, and 2 to the power of 8 is 256 (or, turning that around, the base-2 logarithm of 256 is 8). At first glance, therefore, you might suppose that the entropy of a message encoded in ASCII is just 8 bits per symbol. But that isn't quite right. It doesn't take into account the fact that some combinations of symbols are more likely than others. In English, for example, *u* almost invariably comes after *q*. The calculation of Shannon entropy allows for such factors by multiplying each item of information content by the probability that it will occur. In other words, you take the logarithm to base 2 of the probability of each event, multiply that by the probability, and add these terms together. This way of figuring out the Shannon entropy, known technically as Shannon's noiseless channel coding theorem, gives the shortest message length needed to transmit a given set of data. That is to say, it tells us the optimal data compression–the most compact form of the information–that can be achieved.

It's worthwhile remembering that Shannon called his 1948 paper "A Mathematical Theory of *Communication*" (emphasis added). That was his prime concern: to analyze the ability to send information through a communications channel. So an immediate benefit of his theory was to give engineers the math tools they needed to figure out channel capacity, or how much information can go from *A* to *B* while incurring the least possible number of errors. The information you want is the signal; the extra stuff that gets thrown in along the way is the noise.

Having defined entropy within the new, special context of information, Shannon went on to show how it was related to a channel's information-carrying capacity. He proved that for any channel, you could determine a capacity that, providing it wasn't exceeded, would result in transmissions as free from errors as you cared to specify. There's always a chance of error whenever a message is sent, but by using the tools of information theory you can squeeze that probability down to an arbitrarily small value.

Shannon proved that even in the worst case of a noisy channel with a low bandwidth, it's still possible to achieve nearly perfect, error-free communication. The trick, he found, is to keep the transmission rate within the channel's bandwidth and to use error-correcting schemes. These schemes involve sending extra bits that enable the data to be extracted from the noise-ridden signal. Today everything from modems to music CDs rely on error correction to get their jobs done.

Shannon also showed that much of any real-world communication contains *redundancy*, or message content that isn't strictly necessary. He analyzed a huge number of communications, from code transmissions to phone conversations to James Joyce novels, in order to understand the relationship between the message intended for transmission and the redundant stuff tacked on to ensure that the message is understood correctly. Redundancy is crucial for clear communication in a noisy environment, but when noise is low the redundancy can be stripped out and the message highly compressed.

In this vein, Shannon looked at the frequency of word correlations in the English language. Pairs of words that often appear together (for instance, *straight* and *road*) show a higher degree of redundancy than less common pairs (such as *ugly* and *sunset*). Shannon showed that a randomly generated string of words could sound remarkably like meaningful English, so long as each word had a high correlation with the word before it.

In a classic exposition of information theory, the mathematician Warren Weaver, a close colleague of Shannon's, wrote: "When one meets the concept of entropy in communications theory, he has a right to be rather excited–a right to suspect that one has hold of something that may turn out to be basic and important." Although it wasn't obvious at the time, Shannon had established an important link between the ethereal world of information, which we generally think of as existing in the mind, and the physical world of matter and energy. Hartley and Nyquist had started work on that link two decades earlier, but it was Shannon who took the decisive step of extending the idea of entropy beyond the realm of the material.

Shannon's concept of entropy provides the first clear connection between a fundamental way of measuring information and the physical resources needed to represent information. However, an earlier inkling of this concept is contained in a strange situation dreamed up by James Maxwell (of electromagnetic theory fame) back in 1867.

Maxwell imagined a container that's filled with gas and divided into two halves. A tiny demon stands guard over a small hole in the middle of the dividing partition. His job is to allow only fast-moving particles to pass from the left half to the right half, and only slow-moving molecules to move in the opposite direction. To accomplish this, he has a little shutter that he can rapidly open and close at just the right moment as a gas molecule approaches. The energy consumed in operating the shutter is negligibly small, so it doesn't come into the reckoning.

The upshot of the demon's feverish activity is to cause the right half of the container to heat up because it now contains only high-speed molecules, and the left half to cool down. This represents a massive increase in order and a consequent decrease in entropy. But the demon is working within a closed system. And in such an environment the law of entropy–the second law of thermodynamics–states quite clearly that entropy must always increase, or at least stay the same. Since the second law of thermodynamics is firmly established in physics, something must be missing from the picture of Maxwell's demon, but it isn't obvious what. The resolution of this paradox involves a deep link to the world of information and computers.

As the theory of computers, or computer science, evolved in the second half of the twentieth century, some researchers began to wonder what exactly were the costs in terms of energy of doing an individual computation. Just as an engine inevitably wastes a certain amount of energy, mostly in the form of heat, it seemed reasonable to suppose that, for example, adding two numbers together, which ultimately is done using physical devices like capacitors and transistors, must involve some dissipation of energy. The general assumption was that the minimum amount of heat that was uselessly frittered away per computation would have to be roughly the amount of energy that a randomly moving molecule has at a given temperature.* It was hard to see how any possible calculation could waste less heat than it took to move a single molecule.

But then came a surprising discovery. In 1961, a German-born theoretical physicist working for IBM, Rolf Landauer, proved that a computation *in itself* doesn't have to waste energy at all. The loss comes from a different but related source.

Landauer did more than anyone else to formalize the concepts of information as a physical quantity and computation as a physical process. His chief interest was the extent to which physics constrains computers. This led him into many related areas, including speculation about what future computers would be like. He was one of the first to refer to nanoscale electronics and the important role that quantum effects would play in such Lilliputian devices. Ironically, in the many letters Landauer wrote criticizing what he saw as exaggerated claims about the potential applications of various physics-based technologies, one of his favorite targets was quantum computing–a subject that's now not only respectable but seems likely to flourish, thanks in part to his seminal work on the physics of information.

---

*This energy is on the order of $kT \log 2$, where T is the temperature of the system and k is Boltzmann's constant.

In considering the energy cost of doing a computation, Landauer started from the inescapable fact that to encode information you have to have some kind of physical device. It might be a scrap of paper with figures scribbled on it, an abacus, your brain, a computer. The point is that information can't be disembodied: it has to have a physical support system, and this system, be it brain or calculator, must obey the laws of physics. Therefore information must also obey the laws of physics.

It seems to make perfect sense that in carrying out computations, which involve something changing its physical state, waste heat will be produced somewhere along the line. Landauer didn't question this assumption, but set out to clarify just where the energy loss takes place. And what he found is mildly astonishing: a computation generates waste heat only when information is erased.* This idea became known as Landauer's principle. Rubbing out old results has always played a crucial role in computing because a computational system has to reset itself to some initial state before it can take on a new task. Otherwise, it just fills up with old data and grinds to a halt. That's exactly what happened to Spirit, one of the Mars expedition rovers in 2004: its flash memory, choked full of files accumulated during seven months of flight from Earth followed by several weeks on Mars, made it gradually unusable. For a while, after the probe ground to a halt one Martian night, mission controllers were baffled, but eventually it became clear what had

*Real computers, like the one on your desk, don't operate under perfect conditions and so generate a lot more heat than the ideal system that Landauer considered.

happened. Engineers on Earth started to send instructions to wipe the memory clean, step by step, freeing up room so that the spacecraft could run its programs again and continue with the mission.

Landauer's ideas were taken further by a fellow IBM researcher, Charles Bennett. An amateur musician and photographer with a Ph.D. in molecular dynamics from Harvard, Bennett had joined IBM's Thomas J. Watson Research Laboratory, near New York City, in 1972. He and Landauer had first crossed paths at a statistical physics meeting in Chicago the year before. Landauer recalled of that encounter, "I realized that the number of people in the world interested in a disciplined way in the physics of computation had doubled." At IBM, the two became close colleagues with common interests. "I started more or less as his mentor," remembered Landauer. "Eventually, the relationship inverted, and he became the intellectual leader."

In 1973, Bennett showed that it was possible to design a type of computer that Landauer hadn't considered: a *reversible* computer. Such a machine would first chug away at a calculation to produce some result; but unlike any real computer ever made, it would also keep a precise record of all of the steps it took to reach this answer. Then, having delivered its solution, it would run backward until it returned to its initial state. Working this way, in a kind of closed loop, there'd be no erasure and so, in theory, no waste heat. The same conclusion, reached independently by Ed Fredkin and Tom Toffoli at MIT, proved to be decisively important a few years later when Paul Benioff, at Argonne National Laboratory, used it to show the feasibility of a quantum computer (see chapter 8).

Bennett's work also led him to a remarkable resolution of

the paradox of Maxwell's demon. He realized that in order to perform his trick, the demon must make measurements of the particles in the container so that he can decide which molecules are moving fast and which are moving slowly. In the course of taking these measurements, the demon gains information, which he has to store somewhere–presumably in his memory. But memories have a limited size, even those of demons, so to record new information the guardian of the shutter must eventually erase some of the old data he has collected about particle speeds. As soon as he does this, Landauer's principle comes into play: the erasure causes the demon to dissipate energy in the form of waste heat. Bennett showed that the rise in entropy due to the demon erasing its memory more than offsets the fall in entropy of the gas in the container due to the sorting of molecules between the two halves. Hence, the total entropy for the system always goes up. Far from breaking the second law of thermodynamics, Maxwell's mischievous imp gives us one of the clearest demonstrations of the deep connection between information and the physical world.

Every age has its own set of motifs for understanding the workings of the world. In Newton's era, the notion of a clockwork universe prevailed. The Industrial Revolution was characterized by an obsession with progress, evolution, and ultimate "heath death" when all useful energy would be gone: the cosmos viewed as a kind of giant steam engine. Now we're in the computer and information age, so it's become natural to ask whether the universe itself might not be some kind of colossal information processor.

An extreme advocate of this view is Ed Fredkin, a computer scientist at the Robotics Institute at Carnegie-Mellon University and previously of MIT. Fredkin promotes an idea he calls digital philosophy, according to which the basic elements of the universe are discrete, and therefore can be thought of as digital. Since everything in nature is digital, he reasons, every change from one state to another involves a computation.

Fredkin's claim that all underlying components of the universe, including space and time, ultimately consist of tiny indivisible pieces isn't at odds with modern physical theory. But Fredkin does go out on a limb in asserting that at the basic level the universe is made up of small cells called cellular automata, or miniature computational elements that could assume different digital states. Fredkin imagines that as space-time consists of these elements, it has a certain computational capacity. The universe therefore effectively evolves by computation–by the cellular automata changing from one state to another.

How much truth there is in this paradigm of a cosmic computer is a moot point. Fredkin, in fact, has ventured even further and proposed that the universe, which he maintains is a consequence of information processing, is run by a machine he calls "the other." Here his ideas begin to smack of technical mysticism, and some scientists, such as the MIT physicist Phillip Morrison, view the more radical of Fredkin's speculations as nothing more than a response to the zeitgeist. Yet there's little doubt that the concept of information as something as fundamental as space, time, and matter is already well established in some circles of mainstream science.

After all, everything in the environment we become aware of through our senses is really just information, be it the color of a flower or the warmth of a sandy beach. Life is controlled via the genes in our DNA–a large molecule that's little more than a chemically based information code. Descending further toward the basement of reality, the behavior of the subatomic particles that make up everything in the universe is governed by that mysterious object known as the wave function, which is really a quantification of information. Is an electron over here or over there? In telling us the probabilities of each choice, here or there, the wave function conveys information about the state of the electron. Observations such as these encourage us to look again at the nature of what it is from which everything springs. Is the universe really about quantities such as mass, force, charge, and energy? Or, at the foundational level, does information play an equally important if not primary role?

Central to such considerations is quantum theory. As John Wheeler asks: why the quantum? Why is the atomic and subatomic world made of distinct, indivisible pieces, rather than being continuous the way the macroscopic world of energies, positions, and velocities appears to be?

A recent and intriguing answer has come from the Austrian physicist Anton Zeilinger, one of the world's leading experimental pioneers of teleportation.[52] Zeilinger starts from a position held by Niels Bohr. "It is wrong," said Bohr, "to think that the task of physics is to find out how Nature is. Physics concerns what we can say about Nature." Intrinsic to Bohr's philosophy was the belief that all we can ever know about the physical world is restricted to the information mediated by the senses.

With this link between physics and obtainable information

firmly in mind, Zeilinger begins with a conceptual leap. He assumes that the basic units of information–bits–can be associated with the basic building blocks of the material world. In quantum mechanics, these building blocks are elementary systems, the archetype of which is the spin of an electron. Following Bohr's approach, Zeilinger avoids asking what an elementary system is and focuses instead on what can be said about it. And he makes this very simple but far-reaching claim: *what can be said of an elementary system is contained in one bit of information.* Ask for the spin of an electron and you get the answer "up" or "down"–"zero" or "one." Nothing more can be obtained from the system beyond this one-bit reply; it's like a genie that grants you the answer to one question and then disappears. You might choose to make your measurement of the electron's spin in the *z*-direction and find it to be, say, spin-up. That's all you can know. There are no more bits available to the system for it to say anything about the spin in other directions.

Several basic characteristics of the quantum world follow directly from Zeilinger's principle, starting with an answer to Wheeler's question: why the quantum? Since information has an irreducible graininess to it–there being no smaller piece of it than a single bit–it follows that our experience of nature, which comes through experimental inquiries, will also reveal an ultimate graininess. This is why electrons are restricted to fixed energy levels in atoms, why light comes in chunks we call photons, and why properties such as particle spin and polarization can take only discrete values.

The intrinsic randomness found in the quantum world is illustrated by the example of electron spin we looked at previously. Once you've chosen to measure the spin in a certain direction, the system uses up the only bit it has available in giving you the answer. There's nothing left to say about the spin in the *x*- and *y*-directions, so they're of necessity entirely random.

Extending his scheme to sets of elementary systems, Zeilinger finds that it offers a natural explanation for entanglement. Two elementary systems, he proposes, carry exactly two bits of information, and *N* systems carry *N* bits. When, say, two electrons are entangled, it's impossible even in principle to describe one without the other: each has no independent existence. Zeilinger's principle sees this as a straightforward consequence of the limited information available to the system. The spins of a two-electron system, for example, are represented by two bits. They might be "the spins in the *x* direction are parallel" and "the spins in the *x* direction are antiparallel." This exhausts the information supply, and the state is completely described–yet nothing has been said about the direction of spin of one electron or the other. The entire description consists of relative statements, or correlations. This means that as soon as one spin is measured along a certain direction, the other one is fixed, even if it happens to be far away.

Through a merging of the principles of quantum mechanics and information theory, a new view of the universe is emerging in which the quantization of matter and energy, and perhaps eventually of time and space, too, are seen as consequences of the graininess of information. What we can know about the world at a fine level comes through an elaborate version of the old parlor game twenty questions. In the final analysis, nature must answer yes or no to our most basic questions, and this sets a limit beyond which we begin to see the world not as smooth and continuous but as a composition of pixels.

It's been suggested by some that the universe itself is a computer at the quantum level. Whether this is true, we've certainly built computers that are growing more powerful every year. These machines still process information in a classical way; that is, a bit of information is stored in a physical system as a voltage, a magnetic spin, or a light pulse that can take one of two sharp values, zero and one. However, a new generation of computers is on the horizon that will harness quantum properties, including information represented by a superposition of states.

A single quantum bit, as distinct from a classical bit, can be in two states *at the same time*, so that it can encode two values simultaneously. A two-particle system, such as a pair of interacting photons, thereby gives a superposition of four (2^2) possible states, a three-particle system eight (2^3) states, and so on. A computational approach like this offers extraordinary possibilities, including the ability to carry out vast numbers of calculations simultaneously.

Quantum computers, which we'll return to in chapter 8, will be one of the key applications of teleportation in the foreseeable future. Another will be quantum cryptography, an outgrowth of the age-old study of secret codes.

5

Secret Communications

How can two people in different places swap messages and be sure beyond doubt that no one else has tapped into their communications? That may seem a strange question to ask in a book about teleportation. But although teleportation is a child of the most modern science, its roots lie in the ancient art of cryptography and the quest for an unbreakable code.

For several thousand years, at least as far back as the Babylonians, people have been sending secret information back and forth for military, political, and other purposes. Famously, Julius Caesar is supposed to have used a system (now named, after him, the Caesar cipher), in which every letter in the original message, or plaintext, is shifted by a fixed number of letters. For example, if the shift is by four letters, then every A is replaced by an E, every B by an F, and so on.

The "shift by *n*" rule is a simple cryptosystem and results in a disguised message, or ciphertext, that can be read only by someone who knows the rule. This someone is usually the intended receiver, but it could also be a shadowy third party, or eavesdropper, who has intercepted the message and figured out how to crack the code. Cryptography is the art of creating and using cryptosystems; cryptanalysis is the art of breaking cryptosystems, or seeing through the disguise even when you're not supposed to be able to; cryptology is the combined field of cryptography and cryptanalysis. The information needed to encrypt or decrypt a message is known as the key. One of the big challenges in cryptography is to devise keys that are hard or impossible for an eavesdropper who manages to get hold of a ciphertext to extract. A seasoned cryptanalyst wouldn't take long to see through anything as simple as the Caesar cipher, and even a much more elaborate system fails utterly if both the message and the key fall into enemy hands.

On January 16, 1917, during World War I, the German foreign secretary Arthur Zimmermann sent a telegram in ciphertext to Count Johann von Bernstorff, the German ambassador to the United States. It said that in the event of war with the United States, Mexico should be asked to enter the war as a German ally. Upon emerging victorious, Germany would in return promise to hand back to Mexico the lost territories of Texas, New Mexico, and Arizona. This devastatingly incriminating message was intercepted by British naval intelligence, which held a copy of the diplomatic codebook previously captured from the Germans. The Zimmermann note, as it's become known, consists of a series of four- and five-digit numbers each of which corresponds to a word in the codebook (for example, 67893 = "Mexico").

With a copy of the deciphered telegram in their hands, the British had the means to do what they desperately wanted: persuade the United States to enter the war against Germany. But the British government was reluctant to expose the actual telegram, because the Germans would then immediately suspect their code had been cracked and would change it. The British knew a Mexican decrypted version of the telegram existed, which, if they could get hold of it, they could pretend was obtained through espionage activity in Mexico and not as a result of code breaking. Consequently, the British government contacted one of its agents in Mexico, who managed to obtain a copy of the Mexican version of the telegram. On February 25, 1917, this copy was delivered to President Woodrow Wilson, and on March 1, the U.S. government put the plaintext of the telegram before the press. An initially skeptical American public, which thought the telegram might have been faked, was persuaded by none other than Zimmermann himself, who openly confirmed its authenticity two days after its publication. On April 6, 1917, Congress approved U.S. entry into World War I.

In the same year, a group of electrical engineers working for AT&T in New York was put in charge of finding a secure way to send wartime messages by telegraph. The group discovered that even when multiple messages were speeding along a telegraph wire in both directions, a savvy third party equipped with an oscilloscope could monitor the frequency changes and transcribe the messages. One member of the group, Gilbert Vernam, came up with a method to foil such eavesdropping. He suggested feeding into the machine a paper tape prepunched with random characters. Character by character, the tape is consumed in synchrony with the input message characters. The set of random electrical pulses produced from the tape is combined with the pulses generated from the text in a special way, known as modulo-2 addition, to create an encrypted message. An identical tape at the other end performs another character-by-character, modulo-2 addition, which, by the nature of this arithmetic, cancels out the obscuring characters to reveal the plaintext. Anyone covertly tapping into the message en route, however, would be treated to a meaningless jumble of pulses.

Vernam had shown how messages could be encoded and decoded automatically, on the fly–a crucial breakthrough because it meant that future cryptography wouldn't have to depend on labor intensive, off-line processes. Such a coded message could easily be tagged on to any form of communications, from telephone calls to radio transmissions, and eventually to e-mail messages flashed over the Internet. But in the way it was originally implemented, Vernam's technique had a weakness that was quickly spotted by Joseph Mauborgne, a major in the U.S. Army Signal Corps (and later its chief). Mauborgne found that if the same randomizing tape was used more than once, the coded messages could be partly unscrambled by superimposing them and looking for patterns. To deny an eavesdropper any chance of getting at the hidden content, a unique, random key had to be used for each transmission.

Compared with most cryptosystems, the Vernam cipher is disarmingly simple and so effective that, in time, it became a routine tool of the espionage trade. Suppose Alice and Bob–the names that cryptologists traditionally give to the sender and receiver–are two spies working for the same agency. Each has a copy of the encryption key in the form of a paper pad. These pads are identical, each consisting of pages on which blocks of random letters are written. A block must be at least as long as the plaintext message to ensure that every character can be disguised. To prepare a coded message, the sender, Alice, refers to the random letters on the topmost page of her pad. One way to carry out the encryption and decryption is similar to the Caesar cipher. Suppose the plaintext message starts "Meet Vladimir outside the French embassy at midnight. . ." and the random string on the pad begins "CQTNOX . . ." Alice takes the first letter of the random string, C, equates this to 3 (since C is the third letter of the alphabet) and replaces the first letter of the plaintext by P (three letters on from M). She then takes the second letter from the one-time pad, Q (17), and advances the second letter of the plaintext by 17 places to give V as the second letter in the ciphertext. Each letter in the ciphertext is generated by applying the same rule using one and only one of the random letters on the pad to the corresponding letter of the plaintext. The receiver, Bob, referring to the identical page of his pad, simply reverses the process to extract the original message. Crucial to the security of this system are several factors: the pads must never fall into enemy hands, the letters on the pads must be truly random, and no page or portion of the key must ever be reused for another encryption. From this one-page-per-message golden rule comes the Vernam cipher's more familiar name: the one-time pad.

In the real world, inevitably, mistakes were made and the rules not always followed–with occasionally disastrous results. The reuse of keys by the Soviet Union, owing to a manufacturer's accidental duplication of one-time pad pages, enabled American cryptanalysts to unmask the atomic spy Klaus Fuchs in 1949. During World War II, code crackers at Bletchley Park in Buckinghamshire, England, decoded important transmissions from the German military on a regular basis because the encrypting method used wasn't as foolproof as the designers had intended. The German army high command had asked the Lorenz company to produce a high-security teletypewriter to enable secret radio communications. Lorenz designed a code machine based on the Vernam cipher, but with one significant difference. The problem in a war situation was how to make sure that the same random character tapes were available at each end of a communications link, and that they were both set to the same starting position. Lorenz decided it would be simpler under the circumstances to build a machine to generate the obscuring character sequence. Being a machine–a deterministic device, albeit a fiendishly elaborate one–it couldn't produce a truly random sequence of characters but instead only what's known as a pseudorandom sequence. Unfortunately for the German army, it was more pseudo than random, and this led to its unraveling at the hands of the Bletchley Colossus, an early electronic computer built especially for the job.

Up to this point, no one knew whether there was any way to crack Vernam's cipher if it was used as prescribed–one time only, with true randomness and no intercepts. But in 1949,* in a brilliant paper that swiftly followed his founding of information theory, Claude Shannon showed by watertight

*Shannon made this discovery in 1945 but it was classified at the time.

mathematical logic that, providing all the conditions for security are met, the one-time pad is totally unbreakable.[46] The ciphertext of the one-time pad gives an enemy cryptanalyst no information at all about the message except for its length. On the other hand, Shannon demonstrated, if the key is shorter than the plaintext, portions of the encoded message can always be extracted by someone who mounts a powerful enough analytical attack. Given the provably perfect level of security offered by Vernam's cipher, you might wonder why efforts to develop other systems continued.

The most obvious reason is the length of the one-time pad key, which has to be at least as great as that of the plaintext. In the past, this was far too cumbersome for most purposes. Espionage, yes. Diplomatic communications, including the Washington–Moscow hotline, yes. These high-priority, episodic tasks could realistically take advantage of the only provably secure cryptographic system on the planet. But for routine commercial applications involving huge daily volumes of data, a far more succinct method of keeping information safe from prying eyes was needed. Following the spread of electronic computers, a number of commercial ciphers were developed, the most widely used of which became the Data Encryption Standard (DES), adapted from an IBM cipher in the 1970s. This is based on a secret key of 56 bits (binary digits–ones and zeros) that's used repeatedly over a certain length of time before being changed. The 56-bit key is combined with plaintext, then chopped into 64-bit blocks in a complicated way involving permutations and arcane mathematical functions. Although it lacks the 100 percent security of the one-time pad, it's more than adequate in practice. No quicker method is known for breaking the

DES cipher than to try all 2^{56} values of the key–a hopeless task for a would-be code cracker even using the fastest computers on Earth.

All systems that depend on a single secret, key, however, have an Achilles' heel. Because exactly the same key is used for encoding and decoding, resulting in what's termed a *symmetric* cipher, anyone wanting to use the system legitimately has to be given a copy of the key in advance. This raises the problem of how to make sure that the key remains secret while it's being distributed. In the case of our two spies, for instance, what if the courier delivering a new codebook to Bob in Beijing is bribed en route to reveal its contents to the enemy? Or, what if the other side makes a copy of it while the courier's taking a nap? Then the security of the system is totally compromised and, worse still, everyone but the enemy is oblivious to the fact.

A way around the key distribution problem was discovered in the mid-1970s by Martin Hellman, a mathematics professor at Stanford University in California, and two of his graduate students, Whitfield Diffie and Ralph Merkle.[17] (More recently, it's emerged that a similar technique was developed a bit earlier but kept under wraps at the British government's eavesdropping agency, GCHQ, and possibly also at the National Security Agency in the United States.) Called public-key cryptography (PKC) it uses, instead of a single, long key shared between sender and receiver, two sorts of keys that are complementary and mathematically related. One of these is a public key, which can be shouted from the rooftops if desired; the other is a private key, known only to the receiver. Anyone with the public key can encrypt and send confidential messages, but only someone who has the private key can decrypt and read them. Many of us benefit from PKC everyday, probably without realizing it: it's the basis for most secure transactions, such as credit card payments, on the Internet. In practice, a public-key cipher is usually used in tandem with a symmetric cipher: the public-key cipher ensures the secure exchange of a key for the symmetric cipher, while the symmetric cipher, which works much faster than its public-key counterpart, is used to encrypt the actual message. Because of its crucial role in electronic data security, PKC has been hailed as one of the most significant practical discoveries of twentieth-century math.

Less well-touted but, in the end, perhaps even more far-reaching, was a breakthrough made in 1970 by Stephen Wiesner at Columbia University in New York. Wiesner effectively invented quantum cryptography, an important branch of quantum information science. In a paper entitled "Conjugate Coding," he showed how cryptography could join forces with quantum mechanics to solve two problems in confidentiality that lay beyond the reach of ordinary (classical) physics. The first problem was how to make banknotes that were impossible to counterfeit. The second was how to combine two classical (nonquantum) messages into a single transmission from which the receiver could extract one message or the other but not both. Wiesner's paper was initially rejected–possibly, it's been rumored, because the National Security Agency caught wind of it and quashed the publication. It eventually saw print in 1983,[49] but by then Wiesner's work had been picked up by two other researchers who helped transform his theory into reality: Charles Bennett and Gilles Brassard.

A dozen years younger than Bennett, Brassard was a sharp-thinking mathematician and computer scientist with a keen interest in quantum mechanics. Having earned his

Ph.D. in computer science from Cornell University in 1979, he joined the faculty of the University of Montreal, becoming a professor at the young age of just twenty-four and, subsequently, one of Canada's science superstars. The two first met in October 1979 at a computer conference in Puerto Rico, where Brassard was giving a paper on some new aspect of cryptography, his thesis topic. Bennett introduced himself, and their conversation quickly turned to Wiesner's (as yet unpublished) work on uncopyable banknotes and to the possible implications of quantum mechanics for secure communications. Bennett and Brassard began thinking about how Wiesner's ideas could be combined with public-key cryptography, but then quickly realized something much more interesting: his ideas could be used *in place of* PKC.

The soft underbelly of any private communications channel that relies entirely on classical physics is that it can be tapped, if an eavesdropper is careful enough, without the sender or receiver being aware that there's a spy in their midst. This is because classical physics allows, in principle, any physical property of an object or signal to be measured without disturbing it. Since all information, including a cryptographic key, is encoded in measurable physical properties of some object or signal, there's nothing to stop passive, undetectable wiretapping of a classical key distribution channel.

The same emphatically isn't true in quantum mechanics. In the quantum world, *every* measurement of a system, *every* attempt at monitoring, however subtly done, inevitably changes the system's state. The very act of listening in by a would-be spy causes an unavoidable disturbance that's a potential alarm bell to legitimate users. This makes it possible to design a quantum channel–one that carries signals based on quantum phenomena–that no one can tap without giving his or her presence away.

A number of different approaches to quantum cryptography have been explored over the past couple of decades, but the one first brought to fruition by Bennett and Brassard draws upon Heisenberg's uncertainty principle. It particularly exploits the fact, which we met earlier, that certain pairs of physical properties are complementary in the sense that precisely measuring one property totally randomizes the other, scuttling any chance of learning anything about it. The pair of complementary properties featured in Bennett and Brassard's scheme, as it was originally conceived, involve polarized light.

It turns out in quantum mechanics that certain types of polarization, as in the case of position and momentum, are complementary and therefore subject to the uncertainty principle. For example, rectilinear polarization, in which the polarization is either vertical (0°) or horizontal (90°), is one member of a complementary pair, diagonal polarization (45° and 135°) is the other. The practical upshot is that a measurement can be made to tell between the two states of rectilinear polarization (↕ or ↔) *or* between those of diagonal polarization (⤢ or ⤡); but no measurement is possible that can distinguish between all four states (↕, ↔, ⤢, or ⤡) at the same time. Heisenberg's uncertainty principle absolutely forbids the rectilinear basis and the diagonal basis from being measured simultaneously: observe one and you can't avoid randomizing and losing all track of the other.

Armed with this knowledge, we can begin to explore Bennett and Brassard's quantum cryptographic arrangement. The goal is for the legitimate parties, Alice and Bob, to establish a secret key and be certain that no one else has gained access to it. To achieve this, they use a quantum channel to send signals along and a public (classical) channel, which has a much greater capacity, to discuss and compare the signals.

First Alice sends Bob a series of individual photons, each of which is in one of the four possible rectilinear or diagonal polarization states (↕, ↔, ⤢, or ⤡). She picks these polarizations at random. Likewise, Bob, who has no idea what's going to be coming down the pike, randomly chooses the type of measurement–rectilinear or diagonal–that he applies to each photon. A rectilinear measurement, remember, will correctly diagnose a rectilinearly polarized photon *and* be able to tell whether its polarization is 0° (↕), or 90° (↔), but it's utterly useless for identifying and distinguishing between diagonal photons. If a diagonal photon comes along, a rectilinear measurement will randomize its polarization and have nothing valid to say about what its inbound state may have been; the result it gives amounts to no more than a flip of a coin, with a 50 percent chance of being right. If Bob plumps instead for a diagonal test on a particular photon, he'll be able to correctly tell between ⤢ and ⤡ if it *is* diagonal, but will get no meaningful reading–again, merely a toss-up–if it's a rectilinear photon.

Suppose that Alice sends the 10-photon sequence ⤢, ↕, ↔, ↔, ⤡, ↕, ⤡, ⤢, ↔, and ⤢, and that Bob attempts to measure this in one-to-one correspondence using techniques D, D, R, D, R, R, D, R, D, and D (where R is rectilinear and D is diagonal). Bob secretly records his results, but over the public channel lets Alice know only which measurement types he applied. Alice replies by telling him which measurements were appropriate to the type of photon sent. In the prior example, these are numbers 1, 3, 6, 7, and 10. Alice and Bob then discard all the other results. If no one has eavesdropped on the quantum channel, the remaining results amount to valid secret information that Alice and Bob share. Notice that neither party has disclosed anything over the public channel that an eavesdropping spy could possibly find useful. This is because the shared secret key arises not from any decisions that Alice or Bob made unilaterally but rather from a merging of their random choices.

But what if someone has eavesdropped on the quantum channel? Alice and Bob can check for this by publicly comparing a randomly picked sample of their retained polarization data. Any differences that show up between the two data sets will point the finger at tapping. To appreciate this, suppose that a spy, Eve, has surreptitiously snooped on the photons sent along the quantum channel from Alice to Bob. Like Bob, she's obliged to choose between either a rectilinear or a diagonal measurement for any given photon. If she picks the wrong basis, her measurement randomizes the polarization sent by Alice. Then, even if she passes Bob a substitute photon that's compatible with her measurement, this photon's polarization (which is based purely on the equivalent of a coin toss) has only a 1 in 2 probability of matching the photon that Alice transmitted. If Eve is fortunate enough to pick the correct basis for her intercept measurement, she can slip Bob a photon that's sure to be identical to Alice's. So, the overall effect of Eve's tapping is to introduce errors into one quarter of any data at which she sneaked a peek–a hefty proportion that must give the game away if Alice and Bob do a thorough enough public data review.

If Alice and Bob do find evidence of eavesdropping, they have to scrap all their data and start again with a fresh set of photons. On the other hand, if they don't uncover any suspicious signs, they can feel confident that all of the data that they haven't compared publicly are secret and secure. These data can be converted to binary digits by some agreed-upon convention–say, equating 0°- and 135°-photons to binary ones, and 45°-and 90°-photons to binary zeroes–to form the secret key.

In practice, the method just described for error-checking, although straightforward, is pretty inefficient. Too much information has to be compared and subsequently thrown out to decide reliably whether Alice's and Bob's data sets are identical, allowing for the fact that any eavesdropping might have been spasmodic and therefore have led to relatively few discrepancies. A better approach is for Alice and Bob to compare the parity–the evenness or oddness–of a publicly agreed-upon random sample containing about half the bits in their data. Alice might tell Bob, for instance, that the 2nd, 5th, 8th, 9th, . . . 997th, and 999th of her 1,000 bits of data contained an odd number of ones. If Bob found that the total number of ones in his corresponding items of data was even, he'd know right away there was a difference somewhere. It turns out that if Alice's and Bob's data *are* different, then comparing the parity of a random agreed-upon grab bag of about half the data has a 50 percent chance of detecting this fact, no matter how many errors there are or where they're hiding. If the parity check is repeated twenty times, with twenty different random samples, the chance of missing an error, if there's at least one present, drops to one-half to the twentieth power, or less than one in a million.

All these ideas–the sending, measuring, and reception of complementary polarized photons along a quantum channel; the cleverly arranged, open comparison of results; and the meticulous checking for eavesdropping–are brought together in Bennett and Brassard's protocol for quantum key distribution. For the initials of their names and the fact that they announced it to the world in 1984, they christened it the BB84 protocol.[7] On paper it looked sound, but the acid test would be whether it could be made to work in practice, first in the laboratory and then as the basis of a secure commercial encryption system that governments and companies could use.

Quantum cryptography has a foot in both the computer science and the physics worlds. As with all interdisciplinary subjects, it gains in some ways and loses in others from its dependence on cross-collaboration. One of the snags in getting this new field off the ground has been that computer scientists and physicists do things differently: they have different jargons, different methodologies, even different ways of disseminating new results. In computer science, for instance, it's standard procedure to go public with big breakthroughs at prestigious conferences, whereas in physics it's almost unheard of, and definitely frowned upon, not to publish first in a learned journal. This may seem like a minor detail, but it's one of the reasons that quantum cryptography didn't take off more quickly in the 1980s. In customary style, computer scientists Bennett and Brassard broke the news of their protocol at a conference–the 1984 IEEE International Conference on Computers, Systems, and Signal Processing in Bangalore, India. Few physicists even heard the announcement. Consequently, there was no great rush of experimental scientists around the world trying to turn quantum cryptography theory into practice. However, at IBM's Yorktown facility, where Bennett was employed and where computer science and applied physics lived side by side with a common eye on commercial exploitation, Bennett and Brassard pushed on toward a practical demonstration of BB84.

Brassard, based in Montreal, co-opted two of his students, François Bessette and Louis Salvail, to help on the software side. They wrote the code used by the personal computer that controlled the whole experiment. Separate programs represented longtime cryptographic friends, Alice and Bob, and their eternal adversary, Eve, simulating their actions and collecting their results.

Most of the technical hurdles, however, involved the hardware. This was mainly the responsibility of Bennett and also John Smolin, a recent graduate from MIT, who already knew Bennett through Bennett's stepson, George Dyer, and was looking for something to do over the summer of 1989. As Smolin recalled, "Neither Charles nor I knew much about building anything, but we knew enough to be dangerous." He remembered a time when he was visiting George in Bennett's apartment in Cambridge, Massachusetts:[48]

> Charlie was excited about some new tea he had gotten somewhere. He had set up a little double boiler using a pot and a teapot, explaining how this was the right way to cook the very delicate tea. George and I left the house for some time, returning hours later; when we came into the kitchen, we noticed the teapot. If you know about blackbody radiation, you've probably seen how a blackbody turns invisible in a furnace, radiating the same spectrum as fills the cavity. The situation we found was not quite that, but there was a red teapot sitting in an empty pot on the stove. This would not have been disturbing, except . . . that at room temperature the teapot had been green. . . . The delicate tea had left nothing but

> a faint burnt aroma. . . . [T]hat's the kind of experimental background Charlie and I were starting with.

Considering the challenges ahead, these were less than promising credentials.

Quantum cryptography deals in photons, one at a time. But how do you build equipment that functions at the single photon level? How do you spit out individual light particles in the same way that an automatic pitching machine hurls out one baseball, then another, then another, in orderly succession? There was no technology on Earth that could do that in the late 1980s. Instead of a source that gave out a regular drip of single photons, Bennett and Smolin had to resort to dim pulses of light emanating from a green light-emitting diode, which were filtered and collimated by passing them through a lens and a pinhole. A risk of such an arrangement is that if the flashes aren't dim enough Eve might siphon off a fraction of the light–say, with a half-silvered mirror–read her stolen portion, and let the rest of the flash travel through the mirror to Bob, its polarization undisturbed. Bob might not notice that the signal had been weakened, or he might attribute the weaker signal to natural attenuation in the channel; either way, the eavesdropping would go undetected because the too-bright flash would behave not as a quantum object but as a vulnerable, classical signal. To eliminate this possibility, Bennett and Smolin made sure their flashes were so faint that on average only one in ten actually contained a photon. (A similar technique, using a so-called attenuated coherent source–incredibly feeble laser light–is used in the single-photon double-slit experiment described in chapter 1.) Such photonic frugality makes life harder for Bob, and the system less efficient, because the chance of getting a positive reading from any given flash is much smaller. However, it drastically reduces the chance of Eve poaching photons unnoticed on the side.

The whole apparatus had to be operated in complete darkness, so Bennett and Smolin had the IBM machine shop build an aluminum box, which they spray-painted flat black. Smolin recalls:

> The spray paint never adhered well, and it gave off a bad odor. We worked on it in Charlie's office because we didn't have a lab, so it was always crowded and smelled like paint in there. I doubt this upset the safety people as much as the high-voltage power supply did. One day we drove to a hardware store and bought a 100 VAC switch so that we could at least turn off one of the supplies without climbing under the table to unplug it.

When it was later discovered that some stray light was getting into the box, Bennett and Smolin went out to a fabric store to buy some velvet as a cover, confessing to the bewildered salesperson, when asked, that it was for a quantum cryptography experiment. (Years later, Bennett, who is hair-follically challenged, went outside to lunch with his old colleagues and used the velvet as a sun hat.)

Even with a complete blackout, there was still a problem at the receiving end of the apparatus. The photomultipliers, used as detectors, sometimes produced a dark count–a false response even when no photon had actually arrived. Imaginary conspirators Alice and Bob couldn't keep junking all their data–as they would according to the ideal BB84 protocol if an error turned up–simply because Bob's detectors occasionally recorded a photon that wasn't there. But how could they be sure these errors weren't caused by Eve's illicit bit-watching? And, if they could, how could they correct any hardware-induced errors and carry on building their secret key without giving valuable clues to Eve?

In response to these difficulties, Bennett and Brassard crafted a new mathematical technique called privacy amplification. First Alice and Bob bring standard error-correcting methods to bear, over their public channel, to spot and correct any spurious bits that may have arisen because of imperfections in the equipment. This process could leak information about the data to Eve. But through privacy amplification, Alice and Bob can take their partially secret key, a portion of which Eve may have gleaned, and distill from it a smaller amount of highly secret key, of which it's very unlikely that Eve knows even a single bit.

One evening in 1989, a decade after they began their collaboration, Bennett and Brassard, together with Smolin, ate dinner and drank a bottle of wine provided (along with a corkscrew) by the Canadian. They then gathered in the lab containing the light-tight box, known affectionately from its shape and color as Aunt Martha's coffin, and its electro-optics. With the power on, the software running, and secretly coded photons slipping unseen from Alice to Bob, they successfully ran the first practical demonstration of the BB84 protocol.[6] True, it was a system in miniature, and a fragile one at that: a device, sitting on a lab bench, that sent quantum encrypted signals at the agonizingly slow rate of ten bits per second a grand distance of thirty-two centimeters through open air. But it worked; it showed what could be done and, crucially, it helped convince the scientific community of the merit and feasibility of quantum cryptography, and quantum information processing in general.

Here was a system that offered what no nonquantum technique could: a way for two parties to forge a secret key and be certain that no one had intercepted or tampered with it. Contemporary encryption algorithms ensure that sensitive data stays secret simply because it would take too long–far too long–to work out which mathematical key was used to scramble the information. But they're vulnerable to possible future breakthroughs in cryptanalysis and hardware, including quantum computers, which might crack the algorithms orders of magnitude faster. Even the uncrackable one-time pad, the Vernam cipher, fails if the key itself falls into the wrong hands. In a world becoming increasingly dependent on electronic transactions and the confidentiality of those transactions, quantum cryptography holds the prospect of a radically new approach–a way of distributing keys that is guaranteed to be safe by the laws of quantum physics.

In the wake of Bennett and Brassard's BB84 demonstration, other groups and individuals around the world took up the challenge of driving the technology forward. Paul Townsend and his team at the British Telecommunications Photonics Technology Research Centre (now part of Corning), Jim Franson of Johns Hopkins University, Nicolas Gisin and Hugo Zbinden of the University of Geneva, Richard Hughes of Los Alamos National Laboratory, and others all made important contributions. The thrust of these second-generation experiments was to show how quantum key distribution (QKD) could be taken out of the lab and applied in a commercial setting–reliably, robustly, over useful distances of many kilometers, and at reasonable data rates of thousands of bits per second or more.

In the work that followed, QKD was demonstrated using commercial optical fibers, a vital step toward integrating the technique into existing communications infrastructures such as multiuser local area networks. Better optical fibers and photodetectors were developed to improve the transmission and measurement of quantum signals. In 2001, the Quantum Information Group at Cambridge University's Toshiba laboratory, led by Andrew Shields, unveiled a new light-emitting diode that can be controlled so precisely that it emits a single photon each time it's switched on. Fashioned using standard semiconductor manufacturing methods, it should eventually allow this major component of a QKD system to be built cheaply and quickly in industrial quantities. Two years later, the same group showed a prototype QKD system in action, forming secure keys at a rate of up to two kilobits per second and over optical fibers one hundred kilometers long. The following year, QKD finally emerged from the laboratory into the real world.

In 2004, the American company MagiQ and the Swiss firm ID Quantique became the first to put quantum cryptographic hardware on the market. Customers at the head of the line included governments and armed forces, although a growing number of orders are expected from the private sector. Financial companies and banks will likely use such systems to convey private information between different offices, while Internet service firms are keen to create ultrasecure lines and sell access to them for a premium. The Geneva cantonal government is considering the ID Quantique product for e-voting and other programs. Presently, such arrangements are possible only on the scale of a large metropolitan area. The delicate states of the photons used to represent bits mean that quantum cryptography can only work along unbroken, and therefore relatively short, fiber-optic cables. But a goal for the future is to develop quantum repeaters, analogous to standard repeater technology that allows conventional forms of data traffic to be sent over intercontinental distances. Meanwhile, efforts are continuing for the transmission of quantum keys over greater and greater distances through open air. In 2002, a team of British and German scientists beamed a quantum cryptographic signal in the form of infrared photons between two mountaintops over 23 kilometers apart in the South German Alps. The goal is eventually to extend the range up to at least 1,600 kilometers to allow transmissions back and forth between satellites in low Earth orbits–an important step toward a global key-distribution network.

Around the time Bennett and Brassard were bringing their BB84 protocol to fruition in the lab, a different approach to quantum cryptography was being developed across the Atlantic. Bennett and Brassard's method takes advantage of Heisenberg's uncertainty principle. But at the University of Oxford, David Deutsch, a pioneer of quantum computing, and his Ph.D. student Artur Ekert (who'd also collaborated somewhat with Bennett and Brassard) were hatching an alternative quantum cryptographic system based on entanglement.

One of the weak spots of BB84-type cryptography is that it doesn't offer a secure way of storing the secret key, once it's been established. Alice and Bob need some way to keep a copy or record of the key for use when it's needed. But the longer they hold it–in a safe, on a computer disk, or wherever–the greater the chance that someone will sneak an unauthorized look. This is where an entanglement-based system can help.

Suppose Alice creates a number of entangled photon pairs. She keeps one of each pair for herself and sends the other to Bob. To check if anyone has been eavesdropping, Alice and Bob measure some of their photons right away; the rest they store unmeasured. Just before the key is needed, they measure and compare some of the stored photons. If no one has been surreptitiously peeking at the photons while they've been under lock and key, then Alice will always obtain a one when Bob obtains a zero, and vice versa. If no discrepancies are discovered, Alice and Bob measure the remaining stored photons to extract the key.

Ekert published the first paper on entanglement-based cryptography in 1991.[21] The secure storage aspect of such a system has yet to be demonstrated in practice because it's not possible at present to hold photons for more than a fraction of a second. But the transmission of cryptographic keys by entangled photons has been shown by Paul Kwiat at Los Alamos National Laboratory (now at the University of Illinois), Gisin and Zbinden's group at the University of Geneva, a collaboration led by Anton Zeilinger of the University of Vienna and Harald Weinfurtur of the University of Munich, and others as well.

No commercial quantum cryptography products are yet available that exploit entanglement. But in April 2004, the first electronic transfer of funds using entangled photons was carried out in a bank in Austria. The link, based on a system developed by Zeilinger and colleagues and the Austrian company ARC Seibersdorf Research helped move money between Vienna City Hall and Bank Austria Creditanstalt. The photon-encrypted transfer saw the mayor of Vienna transfer a 3,000-Euro donation into an account belonging to the University of Vienna team. Although the two buildings are just 500 meters apart, a fiber-optic cable had to be fed through 1.5 kilometers of sewage system to make the connection.

In June 2004, BBN Technologies of Cambridge, Massachusetts, announced a further breakthrough: the first computer *network* in which communication is secured with quantum cryptography. Six servers make up the initial configuration of what's called the Quantum Net (Qnet), a project funded by the Pentagon's Defense Advanced Research Projects Agency (DARPA). These servers can be integrated with regular servers and clients on the Internet. Data flows through the network, as between the Austrian sites, by way of ordinary fiber-optic cables stretching the 10 kilometers between BBN's site and Harvard University. Interconnecting multiple nodes was a major new challenge, and this was met by software-controlled optical switches made of lithium niobate crystals that steer photons down the correct optical fiber.

At the dawn of the 1990s, the threads of the story of teleportation were rapidly drawing together. Bennett and Brassard had created the first working quantum cryptographic system. Others were experimenting with entangled states. And the field of quantum information science was beginning to take off. The time was ripe for a remarkable breakthrough.

6

A Miracle in Montreal

A handful of scientists gathered for coffee and conversation in Gilles Brassard's office at the University of Montreal after a seminar in October 1992. A question had been asked at the talk that had drawn them together. They chattered excitedly, drew curious symbols and diagrams on the whiteboard, and swapped a strange blend of ideas from the worlds of quantum mechanics, quantum cryptography, and computer science. As the brainstorming session progressed, there was a growing sense of something new and extraordinary about to emerge. This, they realized, was one of those moments ripe in history for a breakthrough. "And by the following morning," recalled Brassard, "teleportation had been invented."

. . .

Back in the 1980s when Brassard and Charles Bennett were in the early stages of developing their BB84 prototype, interest in the informational aspects of quantum mechanics, though not yet widespread, was steadily on the rise. A number of people from different walks of scientific life were drawn to the field of quantum information and, as happens with like minds, began forming loose collaborative networks.

One of these new recruits was the theoretical physicist Asher Peres, who caught the quantum information bug in 1979 while away from his usual academic base at the Technion Institute in Haifa, Israel, and visiting John Wheeler at the University of Texas at Austin. Wheeler had worked alongside Bohr in the 1930s and was now director of the Center for Theoretical Physics in Austin. Through him, Peres learned of Charles Bennett's work, but it wasn't until 1986 that Peres and Bennett actually met. In the summer of that year, they both had rooms on the second floor of a ramshackle house owned by another quantum informationeer, Tom Toffoli, their host at the Massachusetts Institute of Technology (MIT) where Peres and Bennett were briefly working together.

During a second sabbatical to see Wheeler in 1980, Peres shared an office with a postgraduate student, William Wootters, who had just submitted his Ph.D. thesis, titled "The Acquisition of Information from Quantum Measurements." In this dissertation, Wootters tackled the same kind of scenario that Bennett and Brassard were wrestling with, more or less at the same time, for their cryptographic protocol–two parties using polarized photons to send quantum information back and forth. At the University of Texas at Austin, Wootters had already struck up a collaboration with fellow student Wojciech Zurek, a Polish-born physicist who later moved to Los Alamos National Laboratory. The two shared common interests in quantum mechanics and computing, entropy, and the nature of time.

In 1982, Wootters, now teaching at Williams College in Massachusetts, and Zurek published a landmark paper in the journal *Nature* called "A Single Quantum Cannot Be Cloned."[50] They concluded that it's impossible to make identical copies of a quantum state: perfect replication at the quantum level inevitably leads to the original being destroyed. The no-cloning theorem, as it became known, insists that any attempt to scan the original state, to learn its details, effectively sucks all the information out of it and condemns it to nonexistence. This has profound implications for quantum teleportation. It also seemed at the time that it might be a showstopper for quantum computers because no-cloning bars the usual classical error correction techniques from being applied to quantum states. This means, for example, that backup copies of a state can't be made in the middle of a quantum computation for use in correcting errors that might crop up later. Error correction is going to be vital if quantum computing is ever to become practicable, and for a while it looked as if Wootters and Zurek might have exposed a fatal limitation. But then, in 1994, Peter Shor at MIT and Andrew Steane at the University of Oxford came to the rescue by independently devising quantum error-correcting codes that circumvent the no-cloning theorem.

Wootters and Peres linked up again in 1989 at a workshop on complexity, entropy, and the physics of information organized by the Santa Fe Institute in New Mexico. Wootters was already in residence there, on sabbatical leave from his college, and Peres stayed a couple of extra weeks to go over some ideas with him. They talked about the measurement of quantum systems and singled out one problem in particular for special attention. Suppose, they said, that two quantum systems are in the same state. These systems have been prepared identically and are in different locations. What's the best way for observers Alice and Bob to measure their state? Can more information be gleaned about the common state from a single measurement on the joint system, formed by bringing the systems together? Or, is it more efficient to make separate measurements on each–assisted by ordinary, nonquantum communication between Alice and Bob? Peres suspected the former, Wootters the latter, and after running through some simple examples it seemed as if Wootters might be right. But matters didn't rest there. After Peres went back to Israel, they carried on exploring the problem via BITNet (short for Because It's Time Network), an ancestor of today's Internet. During their long-distance exchanges, they argued back and forth until finally Peres was able to show that a certain type of joint measurement (involving so-called trine states) results in the optimal amount of information. Together they wrote a paper, which was eventually published in 1991.[39]

In October 1992, at a meeting in Dallas on physics and computation, Peres caught up with Charles Bennett and told him of the work he'd been doing with Wootters. Bennett already knew about it: he pulled a copy of the Peres–Wootters paper out of his briefcase and said he'd been showing it to everyone. Later he introduced Peres to Gilles Brassard and the two of them, helped by their common fluency in French, quickly struck up a friendship. Also in Dallas, Peres met two young computer scientists: Frenchman Claude Crépeau and Englishman Richard Jozsa. Crépau had finished his doctorate at MIT in 1990, coauthored a paper with Brassard, and was now doing research at l'Ecole Normale Supérieure, a top academic university in Paris. Jozsa was a student of David Deutsch at Oxford, had also recently earned his Ph.D., and was spending a year working with Brassard's group in Montreal.

Following the Dallas meeting, Brassard got in touch with Wootters and invited him to give a seminar in Montreal later that month. Wootters accepted and flew up to Canada to talk about the work that he and Peres had been collaborating on: the problem of finding the most efficient way for two parties, Alice and Bob, to learn about a quantum state. During the question period after the seminar, Bennett, who was sitting in the audience, raised an interesting point that stemmed directly from research he'd been involved with for the past year or so.

Having successfully demonstrated a working version of their BB84 protocol, which depended on single particles and the uncertainty principle, he and Brassard had gone on to investigate quantum cryptographic schemes that involved entanglement. In fact, Bennett and Brassard, together with David Mermin, a physicist at Cornell University, had just published a paper on a variation of the protocol devised by Artur Ekert that used entangled pairs.[9] So entanglement was at the forefront of Bennett's mind when he asked Wootters: Wouldn't Alice and Bob, in measuring a quantum state, achieve the optimal result if they shared an *entangled* pair of photons?

This was the issue that brought five of the participants at the seminar–Brassard, Bennett, Crépeau, Jozsa, and

Wootters–back to Brassard's office. All five had similar, overlapping interests. Wootters was a physicist, the others primarily computer scientists. But all were at the sharp edge of quantum information research, pioneering the most effective ways to send signals and messages at the photonic level. The ideas they shared in the informal gathering brought them to the brink of a remarkable discovery. After everyone had gone home that evening, Wootters e-mailed Peres, who was back home in Haifa, to say that an intriguing question had come up. By the next day, when the group reconvened, the elements of quantum teleportation had fallen into place.

The six, including Peres, decided to write a paper. This proved an interesting logistical challenge in itself, quite apart from the science of the problem, because the Montreal Six were scattered about in five different places, in four countries (Canada, the United States, France, and Israel), and across eight time zones. "The sun never sets on our collaboration," quipped Bennett, paraphrasing Philip II of Spain.

There was some dispute about terminology. Peres wasn't sure that "teleportation" was quite the mot juste for the effect they were describing. His dictionary (*Webster's New World Dictionary*), defined teleportation as "theoretical transportation of matter through space by converting it into energy and then reconverting it at the terminal point." That wasn't at all what they had in mind, Peres pointed out; however, he was persuaded to go along with the name, but not before Bennett had kidded him that it was all right because he planned to cite Roger Penrose's 1989 controversial book *The Emperor's New Mind* (which briefly discusses teleportation) among the references. For his own part, Peres wanted to say that during the process the quantum state was first "disembodied" and then subsequently "reincarnated." This was overruled, although Peres would later use such colorful metaphors off the record. When asked by a news reporter whether it was possible to teleport not only the body but also the soul, he replied "Only the soul."

A few weeks of intense work and much intercontinental e-mailing brought the paper to completion. Bennett, who served as the chief coordinator and did most of the editing, sent it to *Physical Review Letters* (*PRL*). It was received December 2, 1992. "Alea jacta est" (the die is cast) he wrote to his collaborators, as if like Caesar they'd just crossed the Rubicon. The only fear now was that it would be rejected by the review panel. But the team was pleasantly surprised when it passed, helped by a strong endorsement from one of the referees, who turned out to be none other than David Mermin, Bennett and Brassard's recent coauthor of the cryptography-entanglement paper.

And so teleportation made the jump from science fiction to science fact or, at any rate, to scientific theory. The paper by the Montreal Six–Bennett, Brassard, Crépeau, Jozsa, Peres, and Wootters–crisply and captivatingly titled "Teleporting an Unknown Quantum State via Dual Classical and Einstein-Podolsky-Rosen Channels" appeared in the March 29, 1993, issue of *PRL*.[8] And despite the fact that hardly anyone outside the little community of quantum information researchers who read it had a clue what it was really about, it caught the attention of journalists around the world. That magical, mysterious word "teleportation" had finally been given the scientific seal of approval.

• • •

It's hard to think about teleportation without having images spring to mind of people twinkling in and out of *Star Trek*'s transporters. Not that there's anything wrong with some imaginative speculation about where the subject might lead. A study from Purdue University, carried out in the same year that Bennett and company's paper appeared, found that children learned more about science from *Star Trek* than from anything else outside the home. The "Teleporting an Unknown Quantum State" paper offered no prescription for beaming human beings around, but it did have one outstanding advantage over the famous TV series: it was based on today's science. It showed how something, even if it was just at the subatomic level, could be teleported here and now. The importance of this breakthrough to the scientific research community is underscored by the fact that, according to the *Science Citation Index*, this landmark paper had been cited more than one thousand times by its tenth anniversary in 2003.

Several points need to be grasped about real, quantum teleportation as described by the Montreal Six. The first and most crucial is that it doesn't involve any *thing* traveling through space. What teleports isn't matter or energy or some combination of the two, as happens in *Star Trek*, but *information.* Information about an object is extracted and then sent to another place where the information is used, like a blueprint, to create a perfect copy of the original from local material. The method of extracting the information isn't like anything we're familiar with in the everyday world. Because this information extraction takes place at the quantum level, it so thoroughly disrupts the wave function of the original object that the original loses its identity. If it were a large object, made of many subatomic parts, it would effectively be destroyed by the process. William Wootters and Wojciech Zurek showed there was no way around this in their no-cloning theorem. Losing the original is the bad news about practical teleportation. Gaining an indistinguishable copy in a place where you want it is the good news. And if you're prepared to agree that a perfect replica is just as good as the original, then there are no drawbacks at all.

Teleporting humans would be mind-boggling. Teleporting a mouse, or an amoeba, or even a sugar cube would be amazing. On the other hand, teleporting a single subatomic particle may not seem all that enthralling. The 1993 paper by Bennett and company doesn't even deal with whole particles; it talks about teleporting *states* of particles only, like the spin state of an electron or the polarization state of a photon. Although at first sight this may not fire the imagination, it's an extraordinary breakthrough. Being able to teleport individual quantum states may be pivotal in the future world of quantum computing. And it's an initial step toward teleporting more complex things–such as atoms, molecules, and drops of DNA.

A basic teleportation system involves three parties–say, three friends: Alice, Bob, and Claire. (Eve is now out the picture because no one here is trying to steal secrets.) Claire wants to send Bob, who lives some distance away, a present for his birthday but has left it to the last minute. The only way it will arrive on time is by teleportation. She also doesn't have a lot of money. All she can afford is the polarization state of a single light particle–the direction in which a solitary photon is vibrating. It doesn't seem much but, as she tells herself, it's the thought that counts.

Claire isn't very good at teleporting things, so she asks the more technosavvy Alice to help. Alice can't simply look at

Claire's polarized photon (X) and send the result to Bob because the act of looking–making a direct measurement–would cause a random change (in accordance with the uncertainty principle) so that the measurement result wouldn't be identical to the photon's original state. The key to getting a perfect copy to Bob, as Alice knows, is by not looking, even surreptitiously, but by instead using the weird phenomenon of entanglement. What's needed are two more photons, A and B, that have been created in such a way that they form an entangled pair. One member of the pair, B, is sent directly to Bob, while the other goes to Alice. Alice now takes her entangled photon, A, and combines it with Claire's unseen photon gift, X. To be precise, she measures A and X together in a special way known as a Bell-state measurement. This measurement does two things: it causes X to lose its original quantum state identity, and it also causes an instantaneous change in the entangled photon that Bob has received. Bob's photon alters to correlate with a combination of the result of Alice's measurement and the original state of X. In fact, Bob's photon is now in either *exactly* the same polarization state as the photon that Claire bought for him or in a state that's closely related to it. He doesn't yet know which.

The final step is for Alice to send Bob a message by conventional means, such as a phone call, to tell him the result of her Bell-state measurement. Using this information, Bob can transform his photon so that, if it isn't already, it becomes an exact replica of the original photon X. The transformation he has to apply depends on the outcome of Alice's measurement. There are four possibilities, corresponding to four quantum relations between photons A and X. Which one of these Alice obtains is completely random and has nothing to do with X's original state. Bob therefore doesn't know how to process his photon until he hears from Alice what she found out. He may, for example, have to rotate the polarization through 180 degrees, or he may have to do nothing at all. When he's made whatever change is necessary, he's guaranteed to have a perfect copy of the present that Claire got for him–a photon with exactly the same polarization state as the original *X*.

A few points are worth emphasizing. First, for all practical purposes, photon *B* has *become* the original photon *X*, while *X* itself has been permanently altered (it has effectively lost all memory of the quantum state it started out with) so that it is no longer *X* in any meaningful sense. This is why the effect is called teleportation: it's equivalent to *X* having physically jumped to the new location, even though it hasn't moved materially at all.

Second, *X*'s state has been transferred to Bob without Alice or Bob ever knowing what that state is. In fact, this lack of knowledge is the very reason that teleportation is able to work. Because Alice's measurement of *A* and *X* is completely random, it sidesteps Heisenberg's uncertainty principle; entanglement then primes Bob's half of the entangled photon pair automatically.

Third, teleportation relies on there being two channels or conduits for information–a quantum one and an ordinary or classical one. The quantum channel supports the link between the entangled photon pair and operates instantaneously, as if there were no separation between Bob and Alice. The classical channel carries the information that Alice has to provide for Bob to be able to ensure that his photon is an exact replica of the original *X* or, if it isn't already, that it can be made into an exact replica by a simple operation. The necessity of this classical channel, across which signals can travel only at light-speed or below, means that teleportation takes time even though the entanglement part of it works instantaneously.

Fourth, there are no limits in principle to the distance over which teleportation is effective. An object or property could theoretically be teleported across many light-years. But, again, the process couldn't happen faster than the speed of light and there'd be enormous technical difficulties in making it work over such large distances.

With the blueprints for a teleporter on the table in early 1993, several labs in Europe and the United States began working feverishly to be the first to show teleportation in action. At first glance, the plans look straightforward enough. A handful of photons, some way to detect them–how difficult could it be? But the truth is, building a teleporter posed a major experimental challenge. The problem lay not so much in creating entangled photon pairs; that was something physicists had been doing routinely since the 1980s. The main obstacle to practical teleportation was carrying out the special measurement–the Bell-state measurement–on two independent photons. The first team to do this would be the first to achieve teleportation in the real world.

7

Small Steps and Quantum Leaps

Who built the first car? Or the first calculator? In the race to develop new devices and techniques, there's often a lot of confusion about exactly who crossed the finish line first. Disputes crop up because people disagree on the criteria that have to be met before a winner can be declared. Several machines, for example, have been tagged the "world's first electronic computer," including the British Colossus (1943) and the Small-Scale Experimental Machine (1948), and the American ENIAC (1945), each of which broke new ground in some important respect. But there's no clear-cut winner. In the same way, competing claims have been made about when and where the world's first teleportation took place, with at least four different groups vying for priority.

The Austrian physicist Anton Zeilinger, born in the same month that World War II ended in Europe, doesn't fit the classic stereotype of a white-coated lab geek out of touch with the wider world. "He has an artist's nose," commented one of his colleagues, Peter Zoller, "the artistic intuition of a good painter. He has the brilliant ability to put his finger on the right points, to pick out the raisins in the cake." A bass-playing jazz aficionado with a keen interest in the humanities as well as science, Zeilinger is a short, stocky, bearded, ebullient man. He began his research into the foundations of quantum mechanics in the 1970s. By the early 1990s, his team at the University of Innsbruck was generally reckoned to be among the world's elite in the field of quantum optics.

Nestled deep in the Austrian Alps and best known as a premier ski resort (the 1976 winter Olympics were held here–coincidentally the same year that the summer games were held in Montreal), Innsbruck seems an unlikely place to produce headline-making science. But Zeilinger and his colleagues, Dik Bouwmeester, Harald Weinfurter, Jian-wei Pan, Manfred Eibl and Klaus Mattle, had built up plenty of experience working with entangled photons. The Innsbruck Six were thus ideally placed to take up the challenge of the Montreal Six, laid down in March 1993, and be the first to bring teleportation to life.

Zeilinger's group made entangled photon pairs by a tried and tested method that involved shining light from an ultraviolet laser onto a special kind of crystal, known as a nonlinear optical crystal. The crystal was composed of a substance, such as beta barium borate, that occasionally converts a single ultraviolet (high energy) photon into two infrared photons, *A* and *B*, of lower energy, one polarized vertically, the other horizontally. These photons travel out from the crystal along the surfaces of two cones. If the photons happen to head out along paths where the cones intersect, their wave functions become superposed so that neither has a definite polarization. Instead, their relative polarizations are complementary, which is to say they're entangled. Thinking back to our "birthday surprise" description of teleportation in chapter 6, these photons can serve as the entangled pair *A* and *B*. Measuring one of them to produce a definite polarization value causes its partner to immediately assume the opposite (complementary) state.

Zeilinger and his team also had to introduce the third photon, *X*–the one to be teleported. And, most crucially, they had to carry out that special measurement on photons *A* and *X*, which the Montreal Six had identified. "In Bennett's paper," explained Zeilinger, "there's a single line where they write that a 'Bell-state measurement' has to be made. That had in fact never been done until we did it." Finding a way to carry out this Bell-state measurement was the biggest obstacle that the Innsbruck researchers had to overcome. But by 1997 they knew how to do it.

In the Innsbruck experiment, a short pulse of laser light zips through the barium borate crystal and produces *A* and *B*, the entangled photon pair. *A* travels to Alice, and *B* to Bob–both characters being, in reality, mere detection systems sitting on a laboratory bench. Meanwhile, back in the entanglement box, a mirror reflects the ultraviolet pulse that was used to create *A* and *B* back through the crystal. This generates (sometimes but not always) a second pair of photons, *C* and *D* (which are also entangled, although that fact isn't used for the experiment). Photon *C* immediately enters a detector, the sole purpose of which is to flag that the partner photon *D* is available to be teleported.

D is prepared for teleportation by sending it through a polarizer that can be oriented in any direction. Having emerged from the other side polarized, it becomes our photon *X*, ready to make its mysterious leap across space. *X* is now an independent photon, no longer entangled. And although the *experimenter* knows its polarization, having set the polarizer, Alice isn't privy to that knowledge. *X* continues on its way to Alice and its rendezvous with photon *A*. The apparatus has been finely adjusted to make sure that Alice receives *A* and *X* at the same time: this is the reason that the same ultraviolet pulse was reused.

Next comes the Bell-state measurement and the trick that Zeilinger's group came up with to make this possible. Alice combines *A* and *X* using a half-silvered mirror–a mirror that reflects half of the light that falls on it and transmits the other half. A solitary photon has a fifty-fifty chance of passing through or being reflected. At the quantum level, the phonton goes into a superposition of these two possibilities. In the Innsbruck experiment, matters are arranged so that photons *A* and *Z* hit the mirror from opposite sides, with their paths carefully aligned so that the reflected path of one lies snugly along the transmitted path of the other, and vice versa. At the end of each path sits a detector.

In the normal course of events, the two photons would be reflected independently and there'd be a 50 percent chance of them arriving in separate detectors. But if the photons are indistinguishable and they arrive at the mirror at the very same moment, they interfere with each other. Some possibilities cancel out and don't happen at all; others reinforce and occur more often. Interference leaves the photons with a 25 percent chance of ending up in the same detector. This corresponds to detecting one of the four possible Bell states of the two photons–the "lucky" one that will result in Bob's photon instantly taking on the *exact* polarization state possessed originally by *X*. The remaining 75 percent of the time, the two photons both end up in one detector. This corresponds to the remaining three Bell states. But the way the experiment is set up, there's no way of distinguishing among these.

When Zeilinger and his team carried out their experiment in 1997, they found that in the lucky situation, when Alice simultaneously detects one photon in each of her detectors, they were able to correlate this a large percentage of the time with Bob's photon becoming a perfect copy of Alice's original photon *X*. This correlation was an unmistakable sign that they were witnessing a true teleportation. But the experiment was far from perfect, as the researchers well knew. Working at the level of individual subatomic particles poses extreme difficulties. One of the biggest problems was making *A* and *Z* indistinguishable. Even the timing of when the photons arrived could, in principle, be used to tell the photons apart, so it was important to erase the time information carried by the particles. To do this, the Austrian team exploited a clever idea first suggested by Marek Zukowski of the University of Gdansk. Zukowski's idea involves sending the photons through narrow bandwidth wavelength filters, a process that makes the wavelength of the photons very precise and, by Heisenberg's uncertainty principle, smears out the photons in time.

In the lucky 25 percent of cases, the correct polarization

was detected 80 percent of the time, compared with the 50 percent expected if random photons had been used. The statistics left no doubt that genuine teleportation was happening and being observed.

Quickly, the Innsbruck team wrote a paper, which was published in the journal *Nature* at the end of 1997.[13] Almost overnight, the world's press were on to the story. "Never seen so many photos of Mr. Spock in my life," quipped Zeilinger. That evocative term "teleportation" had piqued the interest of journalists, and, of course, it was the far-fetched possibilities that everyone wanted to read about. When would people be teleporting around as they did on *Star Trek*? "They largely missed the actual innovation of our work," said Zeilinger, referring to the Bell-state measurement.

In fact, the experiment couldn't distinguish between three of the Bell states that it measured–those corresponding to the case where both photons end up in the same detector in Alice's apparatus. Because of this lack of discrimination, 75 percent of the potential teleportation events couldn't be counted, so that the overall hit rate of confirmed teleportations was quite low. Even today no one has managed to devise a complete Bell-state analyzer for independent photons or for any two independently created quantum particles.

Back in 1994, however, Sandu Popescu, then at the University of Cambridge, had suggested a way around this problem.[40] Why not, he said, have the state to be teleported ride shotgun with Alice's auxiliary photon *A* instead of using, as Zeilinger's team had done, a third, distinct "message" photon *X*? With both roles–those of entangled partner and carrier of the state to be teleported–played by the same photon, an ordinary single-particle measurement results in detection of all four possible Bell states. This alternative strategy to teleporting a quantum state was adopted by a group at La Sapienza (the name by which the University in Rome is known) led by Francesco DeMartini.[12]

In the Rome experiment, photon *A* (Alice's auxiliary photon) was split by a semireflecting mirror into two superposed states that were sent to different parts of Alice's apparatus. The two alternatives were linked by entanglement to photon *B* (Bob's auxiliary photon), which had been similarly split. Upon photon *A* being measured in one of its two possible locations with one of two possible polarizations, one of the split alternatives of photon *B* spontaneously assumed the same state, completing the teleportation.

Innsbruck or Rome: which can rightfully claim the teleportation laurels? As far as chronology goes, the Innsbruck team submitted its paper in October 1997 and had it published two months later; DeMartini's group in Rome submitted its paper in July 1997 but had to wait until March 1998 for publication. In each case, the team wrote up and sent in its paper shortly after the experiments were carried out. So in terms of implementation and submission, Rome won. In terms of publishing date, Innsbruck came in ahead.

But more important in deciding priority is what actually counts as an authentic teleportation. On this issue, too, there's disagreement. Rome would like us to believe that its surrogate form of teleportation–having the teleported state piggyback on one of the entangled photons–is bona fide. Moreover, DeMartini has argued that Innsbruck's effort was lacking because it couldn't detect all the Bell states. This meant it couldn't verify that nonlocality, a key feature of teleportation, was involved in its results. Zeilinger, making the case for Innsbruck's priority, has highlighted the main weakness of Rome's claim: although DeMartini and his colleagues detected all four possible Bell states, their method couldn't teleport a *separate* quantum state–one that was associated with a particle that came from outside. In the Rome experiment, the initial photon has to be entangled with the final one at the outset. Zeilinger insists that only the *interference* of independently created photons, as happened in the Innsbruck setup, makes teleportation of an independent and even undefined (not just simply unknown) state possible. As for demonstrating the violation of Bell's inequalities, he responds that this is beside the point because disproving unorthodox alternatives in quantum physics wasn't the aim of the experiment. In the final analysis, both Rome and Innsbruck deserve a place in the annals of teleportation history, with perhaps Innsbruck winning the contest on a technicality.

Speaking of technicalities, though, neither of the European experiments achieved what might be called a complete teleportation in both the letter and the spirit of the scheme proposed by the Montreal Six. Zeilinger and his team were obliged to leave out the last step of the 1993 teleportation protocol. Because they couldn't measure all four Bell states, they left Bob, at the receiving end, unable to complete the final transformation of his teleported photon into the original state of photon *X*. (It would fall precisely into that state only in that "lucky" 25 percent of cases where the Bell-state measurement was possible.) DeMartini and colleagues, on the other hand, while completing all the required stages, came up short as regards the spirit, or generally accepted meaning, of teleportation. The Rome approach couldn't teleport anything outside the system of auxiliary photons.

Both these shortfalls were addressed in a very different

and remarkable approach to teleportation taken in 1998 by Raymond Laflamme at the Los Alamos National Laboratory (LANL) in New Mexico.[33] A native of Quebec City, Laflamme is energetic, personable, and familiar on campus for his quick, long-legged stride between his office and lab at the University of Waterloo, where he's director of the newly founded Institute for Quantum Computing. An unusual blend of experimenter and theoretician, he studied undergraduate physics in Canada before moving to the University of Cambridge for his doctoral studies under the direction of none other than Stephen Hawking. As Hawking records in his book *A Brief History of Time*, Laflamme (along with Don Page at Penn State University) was instrumental in changing Hawking's mind about the reversal of the direction of time in a contracting universe. After a spell at the University of British Columbia and a second stint at Cambridge, Laflamme arrived at LANL in 1994, just after the birth of his son. Toward the end of that year, after learning of the seminal work in quantum error-correcting by Peter Shor and Andrew Steane, he and colleague Emanuel (Manny) Knill laid down the mathematical basis for this subject. A few years later, he took advantage of LANL's cross-disciplinary culture to get some practical experience of nuclear magnetic resonance (NMR). Collaborating with specialists at Los Alamos and MIT, he began putting NMR to work in studies of quantum information science. "The thread in my research," said Laflamme, "has been to understand the limitations, both in theory and in experiments, on the control we have on quantum systems."

As an active field of research, NMR has been around since the 1940s and is probably best known as a medical scanning technique under its alternative name–magnetic resonance imaging, or MRI (this name adopted to avoid scaring patients with the "nuclear" tag). NMR works because the nuclei of some atoms–those with an odd number of neutrons or protons or both, such as ordinary hydrogen and the carbon-13 isotope–have a spin that creates a so-called magnetic moment, which makes the nuclei behave as if they were tiny bar magnets. Placed in a strong magnetic field, each nuclear magnet aligns with the field. But the nuclear spin isn't simple and flat like the spin of a merry-go-round: the thermal motion of the molecule creates a twisting force that causes the magnetic moment to wobble like a child's top. The sample is then bombarded with radio frequency pulses. When the spinning nuclei are hit by the radio waves, they tilt even more, and even flip over sometimes. When the magnetic moment is tilted away from the applied magnetic field, some of the magnetic moment is detectable perpendicular (at 90 degrees) to the applied field.

Different nuclei resonate at different frequencies. For example, a carbon-13 atom has to be hit with a different frequency radio wave than does a hydrogen atom in order to make it flip. Also, similar atoms in different environments, such as a hydrogen attached to an oxygen and a hydrogen attached to a carbon, flip at different frequencies. Conventional NMR looks at which frequencies different nuclei, in different environments, flip in order to determine the structure of the molecule and other of its properties.

Laflamme's goal was to use NMR in reverse–to control the states of nuclei in a known molecular structure with radio pulses so that these states effectively stored information that could be used in quantum computing. The nuclear spin states would represent the data, which could be detected and manipulated using radio waves. Along with colleagues

Manny Knill and Michael Nielsen, Laflamme wanted to implement a quantum circuit–the equivalent of an ordinary electronic circuit at the quantum level. They chose one devised by Gilles Brassard, and hence known as the Brassard circuit, that carries out a simple computation in the same way as does a conventional logic circuit in a computer. It does this through a series of steps that involves a quantum bit being teleported from one place to another. Various parts of the Brassard circuit play the roles of Alice and Bob in the teleportation process, and the overall result is to make a quantum state at the input to Alice's part of the circuit appear as the output in Bob's section. A key difference between the teleportation in the Brassard circuit and the teleportation scheme proposed by the Montreal Six is that it doesn't involve a Bell-state measurement.

Laflamme and his coworkers realized that they could make this circuit work by controlling with NMR the nuclei in an organic molecule. For their experiment, they chose a liquid solution of trichloroethylene, also known as TCE, which has the chemical formula ClCHCCl2. By manipulating with radio waves at different specific frequencies the two carbon atoms (C) and the one hydrogen atom (H) in the TCE molecules, Laflamme's team was able to enact the entire sequence of operations in the Brassard circuit. This meant forcing different parts of the TCE molecule to do the job of four logic gates, then carrying out a quantum measurement to make the quantum spin state of the two carbon nuclei collapse, and finally forcing the action of six further logic gates. In the event, the quantum measurement was carried out by simply letting the system decohere–in other words, allowing the environment to do the measuring.

Laflamme and his colleagues were able to verify that the

quantum spin state of one of the carbon nuclei in the TCE molecule (in Alice's section) had been teleported, as a result of the Brassard circuit, to the hydrogen nucleus (in Bob's part of the molecule). In other words, the experiment achieved teleportation of a nuclear spin over interatomic distances. It didn't suffer the weakness of the Innsbruck experiment in that no Bell-state measurement was needed: the final quantum state was an exact teleported replica of the initial state. It also overcame the weakness of the Rome experiment in that the teleportation involved an independent state. So, for these reasons, the Los Alamos NMR experiment could be called the first "complete" teleportation. It was also the first teleportation to involve material particles–atomic nuclei–rather than photons. On the downside, the method couldn't be applied to a *particular* nucleus because NMR works with a large assemblage of molecules. Also, NMR teleportation can't be extended to work over large distances. It operates only over molecular dimensions. This could make it very useful, as Laflamme showed, as a way of testing logic and memory elements of a quantum computer. But it could never be applied to long-range teleportation.

All the pioneering teleportation experiments mentioned so far–Innsbruck, Rome, and Los Alamos–dealt with *discrete* quantum states, including those of polarization and of nuclear spin. Such states can take on only certain values, like those of a switch that is thrown one way or another. The polarizations, for example, could be vertical or horizontal, the spins up or down. This limitation to discrete states led to claims that a fourth experiment, carried out around the same time, was really the first full-blown teleportation.[22]

These claims were based on the fact that it teleported a *continuous* feature of light.

The fourth experiment came a year after the breakthroughs in Innsbruck and Rome, and was carried out by an international team led by Jeff Kimble, a fifty-something Texan with a dry country drawl who does his practical work in the basement of the California Institute of Technology's Norman Bridge Laboratory–a low-ceilinged, stone-floored cryptlike sanctuary–in Pasadena. This is the subterranean home of Caltech's Quantum Optics Group, which in 1998 joined the elite band of research cooperatives claiming to have achieved teleportation. Kimble and coworkers took a light beam and "squeezed" it, which sounds like a journalist's aphorism but is really a technical term meaning that one of the continuous features of the electromagnetic field associated with light is made extremely precise at the cost of greater randomness in another property. Using a method called squeezed-state entanglement, the Caltech group managed to teleport entire light particles, not just one or two of their quantum states, across a lab bench a distance of about a meter. The technique worked something like holography: the entangled beams playing the role of laser reference beams, and the classical (nonscanned) information that was sent acting as the hologram. The major difference is that instead of merely producing an image of some object, the object itself was reconstructed.

Unique to the Caltech experiment was a third party called "Victor," who verifies various aspects of the protocol carried out by Alice and Bob. It's Victor who generates and sends an input to Alice for teleportation, and who afterward inspects the output from Bob to judge its fidelity to the original input. Kimble described the situation as akin to there being a quantum telephone company managed by Alice and Bob. "Having opened an account with an agreed upon protocol," said Kimble, "a customer (here Victor) utilizes the services of Alice and Bob unconditionally for the teleportation of quantum states without revealing these states to the company. Victor can further perform an independent assessment of the 'quality' of the service provided by Alice and Bob."

For the science of teleportation, it had been a breathtaking twelve months. Four contrasting experiments, two in Europe and two in the United States, had shown that the process worked in a variety of settings, using different types of particles and quantum properties. Not surprisingly, *Science* journal featured practical teleportation among its list of the top ten most significant breakthroughs of 1998.

None of these early teleportation experiments had a range of more than a meter or two–from one side of a lab table to the other. Yet future applications of teleportation, including quantum cryptography and communication between networked quantum computers, will depend on the method being extended to distances of many meters or kilometers. Over the past few years, a lot of progress has been made on this front.

In 2003, Nicolas Gisin, another czar of cutting-edge optoelectronics who practices his science in the windowless underground, smashed the world distance record for teleportation. He and his entourage of a score or so graduate students and research assistants work in the University of Geneva's group of applied physics. In the basement of the university's old medical school building, Gisin oversees a suite of laboratories crisscrossed with laser beams and packed with the usual paraphernalia needed to summon up the magic of entanglement and teleportation: photon counters, interferometers, and plain old mirrors that bounce the lasers around.

These days, Gisin splits his time between academe and business. In 2001, he and three other researchers from the group of applied physics at Geneva, Olivier Guinnard, Grégoire Ribordy, and Hugo Zbinden formed ID Quantique, the company we encountered earlier, which, along with the U.S. firm MagiQ, is a pioneer of commercial quantum cryptography. Sending secret keys in quantum parcels is the first important real-world application of teleportation, and the first that demands that the technique be workable over distances on the scale of a major metropolitan area. Large banks and other finance institutions are potentially big customers.

One of the obstacles to long-range teleportation is that noise in the quantum communications channel causes the entanglement between particles and their properties to come unraveled. The further a photon has to travel through, say, a fiber-optic cable, the worse the degradation becomes. This problem particularly affects polarization, which is the property most often used for teleportation experiments in the lab. For this reason, Gisin and his Geneva team have turned to an alternative approach based on time bins, or short-time windows. The researchers generate photons using ultrabrief laser pulses, count time in these small increments, or bins, and arrange the pulses to occur in specific bins. Entangled photons reside in two time bins at once and in this form are able to survive transmission over fiber-optic lines better than polarization-entangled photons can. The 2003 record-breaking experiment achieved teleportation over a fiber-optic cable two kilometers long, coiled between laboratories that were 55 meters apart.[34]

One year later, researchers from the University of Vienna and the Austrian Academy of Science broke the teleportation distance record again. They used an optical fiber fed through a public sewer system tunnel to connect labs that were 500 meters apart on opposite sides of the Danube River. Not only was the span impressive but the Danube crossing was successfully achieved under real-world conditions, including temperature fluctuations and other environmental factors in the sewer pipe that could potentially have interfered with the process.

In 2002, Gisin's firm ID Quantique sent quantum data very much farther than this–67 kilometers over the fiber-optic cable of the standard phone network linking Geneva and the neighboring Swiss city of Lausanne–though not by teleportation. The following year, British researchers from the Quantum Information Group at the University of Cambridge's Toshiba Lab, led by Andrew Shields, achieved a similar feat over fibers 100 kilometers long. In both cases, the data carried a secure quantum cryptographic key encoded using the BB84 protocol developed by Bennett and Brassard, which doesn't involve entanglement and thus isn't quite as sensitive to degradation. Entanglement is the most fragile feature to try to preserve during transmission, explaining why its range is currently about 50 times less than that of nonentangled quantum data. Many challenges are common to all forms of quantum communication however. Among these is the need to find a method of boosting the signal en route.

When a message of any kind is sent, losses happen along the way. The signal gradually fades and some form of amplification is crucial if the message is to travel a long distance and still be recognizable at the far end. Electronic repeaters based on vacuum tubes invented by Lee De Forest in 1906 were the key to the first U.S. transcontinental phone service; without this technology, even thick cables and loading coils couldn't carry a voice signal more than about 1,500 miles. Today, much of the world's phone traffic travels as pulses of light along fiber-optic cables regularly punctuated by optical repeaters that amplify the message-carrying photon bursts to ensure that they reach their destination.

Ordinary optical repeaters, however, work by making copies of the fading photons. This is fine for the vast bulk of telecommunications traffic that is carried optically, because the signals use classical (nonquantum) properties of light. The same kind of copying can't be done to quantum data without destroying the very quantum states that are carrying the information. This has limited researchers to working with unbroken and relatively short, optical fibers. The 100-kilometer transmission achieved by the Cambridge group was made possible by its development of a new, highly sensitive detector that can pick out the few surviving message photons from the background noise. But the next big leap will involve building quantum repeaters to allow quantum data to be refreshed and sent for hundreds or even thousands of kilometers.[18]

The main challenge is that a quantum repeater must "purify" and refresh photons without performing any kind of measurement, because that would randomize the photons' quantum states. No one has reached the stage of a practical demonstration yet: the first testing of components may be a few years down the road, and the first working quantum repeater is reckoned to be at least a decade away. But a number of ideas are on the table. One possibility is to use atoms to temporarily hold the quantum states of photons before passing them on. In theory, a single atom of an element such as cesium could be entangled with a photon. The atom would be held in a mirrored cavity in which the photon can bounce back and forth many millions of times. Repeated interaction between the atom and the photon would cause the quantum state of the light particle to be transferred to the atom, which would store it for a while before reemitting a fresh photon with the old state. It would be a tricky process to implement, though, because of the difficulty in building cavities that perfectly reflect photons. If the mirrors didn't work with close to 100 percent efficiency, they'd absorb the photon instead of allowing the atom to do so.

In 2003, Alex Kuzmich and his colleagues at Caltech announced that they'd gotten around this problem, in theory at least, by replacing the single atom with a large ensemble of atoms. So instead of the photon interacting with a single atom many millions of times, it would interact once with many millions of atoms. The photon would swap its state with the entire ensemble, and the entire ensemble would collectively become entangled. Happily, the process is reversible so that the photon, having refreshed itself during its brief stay in the atomic community, could be reemitted to continue its journey.[32]

A similar approach to building a quantum repeater has been investigated by a group at Harvard University led by Ronald Walsworth. But instead of working with single photons, the Harvard effort has focused on interactions between pulses of light and rubidium atoms. In 2002, Walsworth's group demonstrated the transfer of quantum states in a laser beam to the atomic spin of rubidium atoms. The rubidium atoms were able to store the spin state of a signal for about a thousandth of a second before having to be refreshed in the same way as a DRAM (dynamic random access memory) in a personal computer. This relatively long time compares with the one millionth of a second or so needed for the storage to take place. The Harvard team's next goal is to demonstrate that quantum data is perfectly preserved during storage and retrieval using their technique, thus creating a prototype quantum repeater. Then the plan is to develop the approach further to form the basis of a quantum memory. Quantum repeaters, in fact, which temporarily hold a fragment of quantum data, are really an intermediate step needed to build such memories, which will be essential ingredients of quantum computers.

At the Jet Propulsion Laboratory, a facility in Pasadena run jointly by NASA and Caltech, researchers have come up with another quantum repeater design based on ordinary optical equipment.[31] It would use elements like mirrors, beam splitters, and photodetectors to purify and transfer entanglement among photon pairs. Entanglement purification makes two or more partially entangled states into one fully entangled state. Entanglement swapping converts entanglement: entanglements between particles *A* and *B* and particles *C* and *D* can be converted to an entanglement between *A* and *D*.

Beam splitters direct photons in one of two directions based on the photons' polarization, and photodetectors at each output of a beam splitter determine a photon's polarization. The repeater is made up of a network of beam splitters and photodetectors that route photons based on whether specific photodetectors detect other photons. The combination of the right paths and detection-triggered routing is enough to carry out entanglement purification and swapping.

To use the system to start a quantum communication, the sender, Alice, would entangle photons *A* and *B*, keep *A*, and send *B* to the receiver, Bob. A repeater in the network between Alice and Bob would generate a new pair of entangled photons, *C* and *D*, and bring together *B* and *C*. This would destroy *B* and *C*, and in the process leave *A* entangled with *D*. The device would then send photon *D* on to Bob, giving Alice and Bob a shared pair of entangled photons.

In another recent development, researchers from Leiden University in the Netherlands have shown how it might be possible to teleport electrons in a metal or a semiconductor. The theory behind this idea was contained in two papers published in 2003 and 2004 and, like Laflamme's NMR work, promises to have an application in quantum computers.[3,4]

In metals and semiconductors, the outer electrons of atoms are only loosely held. If one of these electrons is dislodged, it joins a milling crowd of its unattached brethren in what's known as the Fermi sea. At the same time, the errant electron leaves behind a hole that acts as if it were a positively charged particle. Holes, like electrons, behave as if they were carriers of electrical current, although they move in the opposite direction to electrons. When electrons and holes meet, they cancel each other out.

Teleportation, as we've seen, depends on the preliminary step of creating a pair of entangled particles. Obvious candidates are the electron–hole pairs. But the problem is how to single out and observe individual examples of entanglement and teleportation amid the Fermi sea. This is where the Dutch scientists believe they've made a breakthrough.

To create entanglement in a solid, Carl Beenakker, Marcus Kindermann, and their colleagues propose using electrons added to a region of semiconductor. The entangler would consist of two such electron puddles separated by a thin, leaky barrier. Occasionally an electron would sneak through the barrier, leaving behind an entangled hole with precisely matched properties. The Dutch team suggests drawing on an arrangement of electrons that's been used for other types of experiments in the past. The so-called quantum Hall state, for example, consists of two-dimensional sheets of electrons at low temperatures and high magnetic fields, and is a well probed and understood physical system.

Two of Beenakker and Kindermann's entanglers would be connected so that the electron from one annihilates, or fills in, the hole from the other. The annihilated electron would then disappear into the electron sea, but its quantum state would be preserved in the surviving electron because of the entanglement. If this approach could be made to work, the electron's state wouldn't be teleported far–maybe 100 micrometers at the most. But the method could eventually be among those used to instantly transfer data for quantum information processing.

The ultimate dream of teleportation is to convey complex arrangements of particles, and even large objects, from one place to another. Exciting work along these lines has already been done, as we'll see in chapter 9 before going on to explore some of the extraordinary future possibilities of macroscopic teleportation in the last chapter. But before taking that final plunge into the fantastic, we need to look closely at what promises to be the most important and lucrative application of teleportation over the coming decades. Within twenty years, according to some estimates, prototypes of quantum computers–machines of almost unimaginable potential–may be up and running. And deep within them, a host of particles and quantum states will be silently teleporting back and forth.

8

A Computer without Bounds

There are revolutions, and then there are Revolutions–developments that promise to change everything. As Gilles Brassard has said: "With a quantum computer, made from only a thousand subatomic particles, you could quickly do a calculation that a traditional computer *the size of the Universe* wouldn't be able to carry out before the death of the Sun." The quantum computer, a device with teleportation at its heart, will have extraordinary, almost unbelievable powers. And it may be just around the corner.

The idea of a computer based on quantum mechanics was first explored in the 1970s and early 1980s by physicists and computer scientists who were pondering the ultimate limits of computation. Prominent among these visionaries were Rolf Landauer and Charles Bennett at IBM, Paul Benioff, David Deutsch, and the late Richard Feynman.

Although, in a sense, Landauer is the godfather of quantum computing as he was the first to explore seriously the relevance of quantum mechanics to information processing, he was an outspoken critic of ideas suggesting that a quantum computer might some day be built. Really, it was Paul Benioff, a theoretical physicist at Argonne National Laboratory in Illinois, who opened the door to that possibility in 1980. Prior to Benioff's work, it was thought that because of Heisenberg's uncertainty principle any computer working at the quantum level would dissipate energy and that the resulting waste heat would disrupt the computer's sensitive internal quantum states, rendering it useless. But building on Landauer and Bennett's work, Benioff showed that a quantum computer was theoretically possible in which *no* energy would be dissipated so that it would be free to work in a purely quantum mechanical way.

In 1981, the physicist John Wheeler, who coined the term black hole and had started thinking deeply about information as a fundamental quantity in the universe, threw a party at his Texas home. The guests were all talented young scientists who shared an interest in the foundations of computing. During the course of the evening, Charles Bennett got into a conversation with Israeli-born David Deutsch, a wildly bohemian character at the University of Oxford's Mathematical Institute, and sparked an idea in Deutsch's mind: a quantum computer wasn't merely different–it changed the whole ballgame.

The same realization came at about the same time to the Nobel Prize–winning physicist Richard Feynman, a onetime student of Wheeler's. Could any computer, asked Feynman, built using the principles of classical physics accurately simulate a quantum system? In 1982, he showed that it couldn't–not without suffering an exponential slowdown. Even a fairly simple quantum situation, such as one involving entanglement, couldn't be modeled by even the most powerful classical computer in a time shorter than the age of the universe.

Deutsch, meanwhile, had begun obsessing on the problem of quantum computers. Strangely, he was less interested in the computers themselves than in what their existence would say about the nature of reality and, in particular, about his belief in the many-worlds interpretation of quantum mechanics–the astonishing notion that there are vast numbers of other universes parallel to our own. In 1985, he published a paper that's now generally regarded as a classic in the field.[16] It not only laid down the theoretical foundations for the quantum computer but made plain why such a machine differs fundamentally from the kind of computers we use today. Deutsch showed that, in principle, a quantum computer could perfectly simulate *any* physical process.

The power of a quantum computer stems from the fact that it uses quantum properties, such as the spin of an electron, to represent data, and these properties are subject to superposition. The up or down spin of an electron, for example, can stand for a zero or a one. But, unless measured, an electron exists in a superposition of states, in which it is both up *and* down, zero *and* one, at the same time. Do a calculation using the electron, and you do it simultaneously on the zero and the one–two calculations for the price of one.

Whereas ordinary computers work with bits, quantum computers deal in qubits (pronounced "cue bits"), a term coined at a 1992 meeting in Dallas by Benjamin Schumacher of Kenyon College in Gambier, Ohio. A single electron representing one qubit doesn't seem all that impressive. But add more qubits and the numbers quickly become persuasive. A system of two electrons, representing a pair of qubits, can be in four different states at once (00, 01, 10, and 11), three qubits in eight states, and so on. The increase is exponential: with n qubits, it's possible to carry out a single calculation on 2^n numbers in parallel. A quantum computer built from just thirty-two qubits (in principle, just thirty-two subatomic particles), would exceed the processing power of IBM's Blue Gene–currently the world's most powerful supercomputer with a top speed of about 36 trillion operations per second. If qubits were represented using the quantum states of electrons, all the data on all the world's computers would fit comfortably on a pinhead.

Yet this astounding ability doesn't come trouble free. In fact, in the mid-1980s, just as the power of quantum computing was becoming evident, researchers were in a crisis of doubt about whether it would ever be practicable. One of the reasons for this concern had to do with the fragility of quantum data held in the form of entanglements and superpositions of quantum states. It was clear that ordinary interactions with the environment could quickly degrade and destroy quantum information by the process known as decoherence, and that to counteract this some kind of error correction would be essential. But physicists couldn't see how this would work, because it seemed that detecting and correcting errors would have to involve measuring the state of a quantum system, thereby destroying the very information it contained.

Then in 1994, mathematicians Andrew Steane at the University of Oxford and Peter Shor at AT&T's Bell Laboratories in New Jersey independently discovered workable quantum error-correction algorithms. Like their classical counterparts, these methods are based on building in redundancy, so that the same information is spread over many qubits.

In the same year, Shor made another, far more dramatic discovery–one that set alarm bells ringing around the world and galvanized research in the field. Much of the security of modern cryptographic systems rests on the fact that it's extremely hard to factorize large numbers. To find the prime factors of a 1,000-digit number,* for example, using even the fastest computers available today, would take many billions of years. But Shor proved that just a modest-sized quantum computer, thanks to its massive parallelism, would be able to solve the same problem in a fraction of a second. It would render the majority of codes used for military, diplomatic, and commercial purposes ridiculously vulnerable. "That was the turning point–there is no question about it," said William Wootters. "Many people started working in the field after Peter Shor's discovery." No longer a mere academic curiosity, quantum computing was suddenly on the radars of governments and their defense departments. National laboratories sprouted substantial programs to advance the subject, notably the U.S. National Institute of Standards and Technology in Boulder, Colorado; Los Alamos National Laboratory; and Britain's major military

*A factor is a number that divides evenly into another number. For example, 4 and 6 are factors of 24. A prime factor is a factor that's also a prime number. For example 13 and 17 are prime factors of 221.

lab, the Defence Evaluation and Research Agency in Malvern. Shor's breakthrough prompted computer scientists to begin learning about quantum mechanics, and physicists to start working in computer science. The lure of important applications and also of new physics to be unveiled spawned powerful quantum computing groups at the University of Oxford (arguably the strongest in the world), MIT, Caltech, and a handful of Australian universities. New centers continue to spring up around the globe as a new generation of researchers floods into the field.

The first big step toward building a quantum computer came out of research in the mid-1990s involving nuclear magnetic resonance. Ray Laflamme's NMR teleportation experiment, described in chapter 7, was an early facet of this work. Other pioneering research was done by a team at Harvard and by Neil Gershenfeld (MIT), Isaac Chuang (then at Los Alamos, now at IBM), and Mark Kubinec (University of California at Berkeley). In 1996, these three managed to build a modest 2-qubit quantum computer from a thimbleful of chloroform.

The key idea in NMR computing is that a single molecule can serve as a computer in miniature. Information is stored in the orientation of nuclear spins in the molecule–each nucleus (which can be spin up or spin down) holding one qubit–while the interaction between the nuclear spins, known as spin-spin coupling, serves to mediate logic operations. A combination of strong magnetic and radio-frequency fields is used to control the spin states of hydrogen and carbon atoms in organic molecules. For instance, in a magnetic field with a strength of 9.3 teslas, a carbon-13 nucleus in a chloroform molecule wobbles, or precesses, at about 100 MHz. By zapping the molecule with radio waves at this and other precise frequencies, it's possible to address and manipulate individual nuclei in a specific order to carry out logic operations. Effectively, the series of radio pulses acts as the software, while the liquid chloroform (or other organic substance) serves as the hardware.

Gershenfeld, Chuang, and Kubinec used their 2-qubit computer to run a sorting algorithm (a step-by-step procedure for putting items in consecutive order) that had just been discovered by Lov Glover of Bell Labs. Like the factorizing algorithm found by Shor two years earlier, Glover's method revealed an application in which a quantum computer was light-years ahead of its classical counterpart. The 2-qubit chloroform device used the procedure to find a marked item in a list of four possibilities in a single step (compared with two or three steps that would be needed by a conventional computer).

Since then, researchers at Los Alamos, IBM, and the University of Oxford have built 3-qubit (1999), 5-qubit (2000), and 7-qubit (2001) computers based on NMR technology. In 2002, Isaac Chuang managed to coax IBM's 7-qubit computer into carrying out a simple factorization (of the number 15 into its prime factors 5 and 3) using Shor's algorithm. Researchers at Oxford are currently experimenting with the DNA base cytosine.

Unfortunately, quantum computers based on liquid NMR aren't likely to get much more powerful than they are now. The readout signals they produce weaken exponentially with the number of qubits involved in the calculation, because the proportion of molecules found in the appropriate starting state decreases. Thermal interference from the surroundings is the culprit. As a result, scientists don't expect to be able to handle any more than about a dozen qubits by this method before the signal becomes indistinguishable from the background. Attempts to build liquid NMR devices that manipulate ten or more qubits continue, but the long-term future of quantum computing probably lies elsewhere.

While liquid NMR seems likely to run out of steam at some point, research has been gathering pace into an NMR-like technique that would work with single atoms in a solid. First suggested by Bruce Kane at the University of Maryland in 1998, the idea is to bury an array of phosphorus atoms in silicon and overlay it with an insulating layer. On top of this sits a matching array of electrodes, each of which can apply a voltage to the phosphorus atom directly beneath it.

As in NMR, the spin of the nuclei can be flipped by applying radio waves at just the right frequency. Normally this would flip every nucleus, but phosphorus atoms are unusual in having a single electron in their outer shell that interacts with the nuclear spin in a complex way. Applying a voltage to the atom changes the energy needed to address both the nuclear and the electronic spin, and so changes the frequency of the radio waves needed to flip the nucleus. This means that by applying a voltage to a specific electrode and zapping the array with the new frequency, it's possible to target a single nucleus.

A key element of any quantum computer is a special kind of logic gate known as a controlled-NOT that acts on two entangled qubits. To enable this essential 2-qubit operation, Kane suggested applying voltages between adjacent phosphorus atoms in the array to turn on and off the interactions between the outer electrons in each atom.

One of the big attractions of Kane's scheme is that, unlike many others in quantum computing, it's scalable. In other words, it can be expanded indefinitely by tacking on extra qubits. The reason for this is that the qubits can be individually addressed by electrical means. Building a working prototype has become a big focus of quantum computing research in Australia, especially at the Centre for Quantum Computer Technology at the University of New South Wales, under the guidance of Robert Clark.

The strange phenomenon of superconductivity may also prove useful in building quantum computers. In 1999, at the Delft University of Technology in the Netherlands, a team designed a superconducting circuit in which superposed counterrotating currents serve to store and manipulate qubits. The circuit consists of a loop with three or four Josephson junctions for measuring the circuit's state. (A Josephson junction is itself a circuit, made of two superconductors separated by a thin nonsuperconducting layer, that's able to switch at high speed when run at temperatures close to absolute zero.) The fact that the Delft circuit is made by conventional electron-beam lithographic techniques makes it particularly amenable to large-scale integration. But there are a couple of big problems to be overcome: superconducting circuits have very short decoherence times, and today's techniques for measuring the states of the circuits are too invasive to usefully manipulate qubits.

Another solid-state approach involves so-called quantum dots–tiny electron traps on the order of a micrometer or even nanometers in size that are manufactured on a piece of silicon. Each dot may contain anywhere from a single electron to many thousands of electrons; qubits are represented by the spin of these trapped particles. Studies of quantum dots began in the early 1990s when it was found that the trapped electrons behave like artificial atoms, with their own equivalent of a periodic table and pseudochemistry. Then in 1998, David DiVincenzo of IBM and Daniel Loss of the University of Basel in Switzerland suggested using quantum dots as the building blocks of a quantum computer. Since then, various ideas have been touted for exploiting the dots' quantum properties to this end.

One idea is for a 2-qubit system consisting of two electrons shared by four quantum dots in a square. The electrons, seeking to minimize their energy, take up position at opposite corners of the square, and since this arrangement has two configurations, they exist as a superposition that can be manipulated through electrodes at the corners of the square. A number of other techniques involve reading and writing data to the dots with laser pulses and placing a single nucleus at the center of each dot that can be addressed with NMR techniques similar to those used in Kane's proposal.

Yet another potential basis for quantum computers was first suggested in 1995 by Ignacio Cirac and Peter Zoller of the University of Innsbruck. Since then it's become a real front-runner in the field and been responsible for some major breakthroughs. Cirac and Zoller suggested building quantum logic gates from ion traps–devices that were already much used in spectroscopy. The idea is that a number of ultracold ions are held in a device called a linear radio-frequency (RF)

Paul trap. This sets up a high-frequency RF field that restrains the ions tightly in two dimensions but only weakly in the third (up-and-down) dimension. Because the ions all carry like (positive) charges, they push each other apart and tend to arrange themselves equally spaced in a straight line, like beads on an elastic string. This spacing under tension allows them to vibrate as a group in ways that are useful for quantum computing.

The qubits are initially stored in the internal spin states of the ions relative to a background magnetic field and are written to the ions using a pulsed, oscillating magnetic field. A long pulse flips the bits from spin-up to spin-down, or vice versa; a pulse half as long puts them in a superposition of up and down states. An advantage of ion traps is that this superposition is extremely robust, which gives plenty of time to carry out whatever logic operations need to be done.

To share the qubits between the ions, researchers turn to the ion vibrations. The aim is to chill the ions until as a group they're absolutely still; this is then the ground state of the system. Inject a little energy, and the ions begin to jiggle. But being quantum particles, the ions can exist in a superposition of the ground state and the vibratory state, so the vibration can be used to store a qubit. Because the ions all take part in the vibration, this qubit is shared among them. It's as if their collective motion is a kind of data bus, allowing all the ions to temporarily have access to the information and become entangled. This sharing enables the IF- and THEN-type operations, which are the elements of computer logic gates. For example, an instruction might be: IF the vibrational state is 1, THEN flip the qubit in the first ion's internal spin state.

Researchers led by David Wineland at the National Institute of Science and Technology (NIST) have already shown that a string of four ions can be entangled, and are optimistic that more complicated entanglements are in the cards.

At least five groups around the world are working on ion trap quantum computers, with Wineland's team at NIST widely regarded as the front-runner. His group has built a 2-qubit logic gate using a single beryllium ion cooled to its vibrating ground state. With a laser focused on the ion, the group superimposed on the background magnetic field a second magnetic field with a magnitude that varied with the position of the ion. The ion's vibration caused it to experience an oscillating magnetic field, and when the frequency of the oscillation matched the energy difference between the ion's two spin states, energy was transferred from the spin to the vibrational state, mapping the quantum information to the vibration from the spin state. This is the basis of a controlled-NOT gate, and was realized in 1995 only a few months after Cirac and Zoller's original suggestion of the technique. Reading the data involved scattering light off the ion, since a spin-up ion can be made to scatter strongly, while a spin-down ion will scatter hardly at all.

Despite their promise, ion traps have their limitations as well. One is the short decoherence time of the qubits after transfer to the vibrational data bus. Because the ions are charged, the vibrations are strongly affected by any stray electric fields. Nevertheless, researchers are upbeat that this tendency can be overcome by better isolating the trap from the environment. Ion traps also suffer from problems of scalability. As more ions are added to the trap, the risk grows of setting off uncontrollable vibrational states and so destroying the calculation. The next step will be to build adjacent traps, each holding only a few ions, and try to send quantum information from one trap to its neighbor. As we'll see in the next chapter, ion traps are featured in the biggest breakthrough to date in the prototype engineering of quantum computers and teleportation.

Common to all these technological blueprints for a quantum computer are a number of challenges that need to be overcome. Among the most outstanding is the one just mentioned, decoherence: the tendency of quantum states, including the superposed ghost states representing qubits, to be broken up by interaction with the environment. Decoherence is both a bane and a necessity of quantum computing. Natural decoherence is a huge problem because researchers don't want qubits degrading and falling apart while a computation is in progress. This would be the equivalent of losing bits altogether in a conventional computer. On the other hand, intentional decoherence, otherwise known as measurement, is essential to getting the results out of the computer: at some point, you have to force the wave functions of qubits to collapse in order to produce a definite answer.

The general modus operandi of a quantum computer is for it to work well isolated from its surroundings so that the qubits, which are in various states of superposition and entanglement with each other, can continue the unhindered computation of which they're a part. Even the slightest interactions with the outside world could have unwanted effects or ruin the results altogether. Such interactions might include, for instance, accidental entanglement with an outside particle, thermal interactions, or even a stray cosmic ray passing through the apparatus. Any of these factors could upset the delicate internal quantum states within the computer and effectively produce an unwanted measurement, causing one or more qubits to collapse prematurely from their superposed states to definite values, zeroes or ones, before the calculation is complete.

The effort to combat such unwanted effects has spawned several areas of intense study, including those of quantum noise and quantum error correction. When planning strategies for overcoming errors in a quantum system, a major factor to bear in mind is that quantum states can't be cloned. In other words, we're not able to make copies of unknown data states to keep the error rate down. Another difficulty centers again around the measurement problem. To correct errors, you have to know what the state of a system is. But to know what the state of a system is you have to measure it, and measuring a quantum system changes the state and destroys superposition. These types of problems have been tackled using a two-step strategy of measurement followed by a recovery process, allowing error correction to be carried out. But it isn't clear that error correction alone is enough to get over the stability problems of quantum computing.

Until a qubit interacts with the macroscopic world and the well-known rules of classical physics, it behaves according to the laws of quantum mechanics, which are also well understood. However, it's that junction between the classical and quantum realms, in which decoherence occurs, where our knowledge is hazy. Ever since quantum theory came of age in the 1920s, most physicists have been content to gloss over the goings-on at this mysterious interface between the microcosmos and the macrocosmos. It's left us with a long-unexplained gap at the heart of modern science. But if quantum computers are ever to be realized, that void in our understanding has to be filled. In fact, one of the greatest benefits of quantum computing to date is that it's provided a powerful incentive for physicists to probe, experimentally and theoretically, areas that have long been neglected. This new research is giving us a much clearer window into how things really work at the quantum level and into the region where the quantum and the classical collide. Practical, hands-on experience, in particular, is making the subatomic world seem a little less intractable.

Researchers are finding that there's much more to the process of decoherence than was previously suspected. Contrary to the old idea of instantaneous wave function collapse, decoherence isn't an all-or-nothing affair: it happens gradually, albeit on a short timescale, and can actually be quantified. Investigations by Wojciech Zurek and his colleagues at Los Alamos have shown that the rate at which decoherence occurs can be measured by something called the Loschmidt echo. This effect, named after a nineteenth-century German physicist, is an observable measure of the sensitivity of a quantum system to changes in the energy of the system.

Decoherence can also be actively discouraged–if you know how. Although the naive expectation is that any interaction between the qubits of a quantum system and the outside world will provoke decoherence, it turns out that the right kind of external signals can in fact *prolong* the period of coherence. Chikako Uchiyama of Yamanashi University in Japan and others have shown how, in the general case, the application of very short pulses, evocatively named bang-bang pulses, at regular intervals can serve not only to suppress decoherence but also to maintain entanglement–the quantum coupling between several qubits that allows computations to get done. In fact, in the absence of such pulses, disentanglement happens even faster than decoherence, so there's even more of a need to suppress it. The specific form of the pulses, Uchiyama says, depends on the quantum-computing technology in question; in NMR, the pulses could be of the magnetic field, while for quantum dots, the electric field would be pulsed.

Such findings may ultimately lead to quantum computers in which we have pretty good control over the level of decoherence. Even so, there comes a point at which we have to reach into the system and ask it for a definite result.

David Deutsch was the first to show that the power of the quantum computer lies in the way qubits evolve over time. Time evolution in quantum mechanics takes place via the Schrödinger equation, and what this allows is the rapid solution of special kinds of problems–those for which the result depends on all the answers taken together rather than on one specific answer. Although this limits the range of applications that can be usefully tackled to those suited to a massively parallel approach, it also provides a class of problems where quantum computing is exponentially better than its classical cousin.

The difficulty comes when we want to prod the machine to tell us what it's come up with. Somehow, despite the fact that its superposed qubits are exploring untold numbers of alternative possible solutions, we at some point have to exorcise the ghosts and force the quantum computer to converge on a clear solution. But how do we do that and get the answer we want–the right one? It seems that by making a measurement we'll cause the superpositions to collapse to a random state, and then nothing will have been gained.

Peter Shor showed the way around this conundrum in

1994 with his first practical algorithm for a quantum computer. The trick is to allow all the components of the final superposition to interfere with each other. Each of the superposed states has a probability associated with it that has a wavelike behavior: it can interfere with the probabilities of other states destructively or constructively, just as the overlapping waves in the two-slit experiment interfere. Getting the desired answer to a calculation means processing the information in such a way that undesired solutions interfere destructively, leaving only the wanted state, or a few more or less wanted states, at the end. A final measurement then gives the hoped-for solution, or in the case of a few final states, a series of measurements gives their probability distribution from which the hoped-for solution can be calculated.

Another problem facing many of the technological approaches to quantum computing–NMR, ion trap, superconducting, and so on–is scalability. With a few exceptions, the schemes show promise at the small, prototype level but quickly run into difficulties as the number of qubits increases to the level that would be practically useful. A powerful stand-alone quantum computer would need to be built from hundreds or even thousands of qubits. This has persuaded many scientists that if quantum computing is to become an effective force anytime soon, it will have to involve networking small quantum computers together. But sending quantum information from one place to another is tricky. One option is to physically move the qubits, but then they're liable to fall apart through decoherence, and in any case such physical movement would slow down the computation. A much faster and more secure approach to quibit transfer is likely to involve teleportation.

In fact, teleportation will almost certainly be vital to the success of quantum computing at several levels. Once the interfacing problems have been overcome, long-distance quantum teleportation will be crucial to distributed quantum information processing and for networking between different nodes within a quantum computer. Short-distance teleportation, on the other hand, will play a pivotal role in keeping track of information inside quantum computers, moving qubits between logic gates, and other internal tasks.

Teleportation also circumvents the inherent problem raised by the no-cloning theorem that quantum data can't be copied without being destroyed, and it offers a more straightforward way of ensuring data integrity than does error-correcting code. The Brassard circuit, described in chapter 7, is among the options available for this kind of short-range teleportation of quantum data. Such circuits would take the place of electrical connections in a conventional, classical device. So in this computational role, we can think of teleportation as acting as a kind of invisible quantum wiring.

In 1999, Isaac Chuang (IBM) and Daniel Gottesman (Microsoft) also described a way in which teleportation could be used to make "quantum software." The object of this would be to give fairly small quantum computers more power and allow them to run more reliably. Teleportation would be used to prepare combinations of various quantum states that could carry out basic quantum computing operations analogous to the elementary gates (such as the AND and OR functions) and integrated circuits found in today's conventional computers. These prepared states could be stored and then delivered to a quantum computer when needed, so that a computer working with relatively few qubits would be able to carry out complex operations at speeds that would otherwise demand a far larger machine.

Quantum software could become very big business. One possible scenario is that small quantum computers will eventually become as common as conventional PCs are today, but running them and producing reliable results will be more problematic. A solution will be for users to acquire from a vendor quantum states that are difficult or inconvenient to prepare. The proprietary quantum program set up to perform some valuable, specific task would be churned out by the manufacturer in multiple copies using some special-purpose device, tested to assure quality, and stored until needed. The consumer would then download the quantum state for a fee and plug it into his or her hardware, which would act on the state according to some standard protocol.

The biggest advantage of this approach from the user's standpoint will be to drastically cut down the number of errors during a computation. Quantum computers work by executing a series of quantum gates, but each of these is fragile–some types more so than others. Implementing the more sensitive gates in software rather than hardware means that they can be thoroughly checked out in advance and rejected if found to be flawed. The user then knows that he or she is getting a quantum state that is reliable and won't disrupt the computation partway through.

Chuang and Gottesman's central idea is that applying a quantum software routine can be seen as a generalized form of teleportation. In standard teleportation, Alice destroys an unknown quantum state and then sends a classical message to Bob who proceeds to reconstruct a perfect replica of the original state. Chuang and Gottesman realized that if the software is suitably modified, the reconstructed quantum state is modified, too. Instead of winding up with Alice's original state, Bob reconstructs a state to which a quantum gate has been applied. If Alice and Bob use different software, they execute a different gate.

Unfortunately for the user, but to the delight of the vendor, quantum software is a consumable product: once applied, the quantum state is destroyed. Little wonder that one of the researchers involved in this work happens to be employed now by Microsoft.

Transmitting software to quantum computer users would be one of many applications for a future "quantum Internet," a term coined by researchers at Los Alamos. Linking quantum computers together to enable quantum communication over remote distances, distributed quantum computing, and teleportation will almost certainly involve entangled photons flying along optical data highways. But the devil is in the details–especially the details of how to get quantum data out of the memory of one quantum computer, along a quantum communication channel, and into another quantum memory without losing or altering qubits along the way.

In 2000, Seth Lloyd and Selim Shahriar at MIT and Philip Hemmer at the U.S. Air Force Research Laboratory in Lincoln, Massachusetts, suggested sending entangled photons over optical fibers to nodes containing cold atoms that would absorb the photons and so store the entanglement. A number of groups are working on this idea, including those led by Jeff Kimble at Caltech and Eli Yablonovitch at the University of California at Los Angeles. In 2004, Boris Blinov and colleagues at the University of Michigan, Ann Arbor, announced a major breakthrough toward developing a prototype quantum Internet: the first direct observation of entanglement between stationary and "flying" qubits.[10]

The Michigan group is pursuing the ion trap approach to quantum computing in which each of the computer's qubits is stored in an ion's energy state. The way ahead with this technology will involve gradually introducing more ions and finding some way to wire them together. What Blinov and his team demonstrated is that hitting a trapped ion with a laser pulse makes it emit a photon carrying an exact copy of the qubit's information. If the information in the photon is manipulated, this change is transmitted back to the ion. The laws of quantum physics effectively wire the photon to the ion, opening up the possibility of distant ions being able to interact by sending each other photons.

In a separate development reported in 2001, two teams of scientists achieved another remarkable feat bringing the quantum Internet a step closer. They managed to do what seems impossible: stop a light beam in its tracks. Lene Hau's group at Harvard University and Ronald Walsworth's group at the Harvard–Smithsonian Center for Astrophysics, both in Cambridge, Massachusetts, did separate experiments involving shining a pulse of light into a chamber of gas in which the beam got slower and slower and dimmer and dimmer before coming to a complete halt. Hau's group used sodium gas, chilled to within a few millionths of a degree of absolute zero, as a brake, while Walsworth's team used gaseous rubidium. These gases are normally opaque to light, but they can be made transparent by illuminating them with a laser beam called the coupling beam, thereby allowing a "probe" laser pulse to pass through. The process is dubbed electromagnetically induced transparency. If the coupling laser is turned off while the probe pulse is in the gas cloud, the probe comes to a full stop. Basically, the information in the photons is transferred to the collective spin of the gas atoms. If the coupling beam is later turned back on, the probe pulse emerges intact, traveling at its usual speed of 300,000 kilometers per second, just as if it had been waiting to resume its journey.

Quantum computing specialists weren't slow to spot the significance of this result. Sending a photon from one place to another, as in a quantum Internet, is the easy part. But capturing it at the other end is a real challenge. The sodium and rubidium gas parachutes show that ways are available to catch photonic qubits on the wing and hold them a while before letting them fly on.

Quantum computers open up a wonderland of possibilities. Beyond performing the routine tasks of sorting, database searching, cryptographic analysis, and the rest, they'll serve as an astonishing portal into the behavior of microscopic systems, with applications in chemistry, drug design, fundamental particle physics, and nanotechnology. A quantum computer simulation would act as a powerful laboratory able to study nature on the smallest of scales.

Seth Lloyd and Daniel Abrams at MIT were among the first, in 1997, to examine in detail how such simulations would become feasible. They described how a quantum computer could be used to simulate a system of many particles, such as a group of electrons. The number of variables involved in modeling this type of system lies far beyond the scope of any conventional computer. To model a group of one hundred electrons would require a computer to store 2^{100} complex numbers, each with a real and imaginary part, to model the state of the spins. Storing this many numbers in a computer memory is problematic enough, but it's also necessary to model the evolution of the system in time, something that would require two-dimensional arrays with $2^{100} \times 2^{100}$ elements. Yet using a quantum computer, such a simulation could be handled quickly and easily using just one hundred qubits.

Simple proof-of-concept quantum computers, working with a handful of qubits, are already up and running. To extend the technology to deal with 10, 20, 100 qubits, and more, will test the ingenuity of researchers over the coming decades. But few specialists in the field doubt that practical quantum computing will be achieved well within this century. When it is, teleportation, working both within and between these extraordinary machines, will begin contributing to a change in the world as we know it. And that may be just the beginning.

9

Atoms, Molecules, Microbes . . .

The following amazing story leaked out on the Internet in 2001:

> The Department of Defense announced Monday that research scientists at the Massachusetts Institution of Technology have successfully tested the first quantum teleportation. Two white mice, weighing between 87 and 90 grams each, received clear bills of health after they were simultaneously converted to photons of light and then transported 13.7 meters through a hydrogen gas tube. They were interstitially reconstituted within 10 seconds and exhibited physical movement 17 seconds later.

Surprisingly the major newswires and networks failed to pick up on this astounding breakthrough. Perhaps it had something to do with the dateline, April 1. Or maybe it was MIT being called an "Institution" rather than an Institute that gave the game away.

Let's face it, we're all suckers when it comes to the fantastic notion of teleporting living things. Everyone wants to know when it's going to be possible to achieve human teleportation. But there are one or two little problems that need sorting out before anyone can realistically expect to beam up. In fact, there are about 7,000 trillion trillion problems, because that's roughly the number of atoms (99 percent of which are hydrogen, oxygen, or carbon) that make up a human body weighing 70 kilograms (154 pounds). Inside atoms are electrons, protons, and neutrons, so the total number of interacting particles inside each one of us is even more tremendous–on the order of 26,000 trillion trillion for every 70 kilograms of body mass. How could the instantaneous quantum states of so many specks of matter be made to dematerialize and reappear perfectly in a different place?

Scientists have set more modest goals for the foreseeable future. They'd be happy to be able to teleport a single atomic nucleus, or perhaps a whole atom or pair of joined atoms. About the most bullish they get in public is reflected in a comment made by Anton Zeilinger in a 2000 *Scientific American* article: "The entanglement of molecules and then their teleportation may reasonably be expected within the next decade. What happens beyond is anybody's guess."

• • •

Several major roadblocks lie in the way of teleporting large or complex systems, such as a glass of water or even a molecule of water. First, the object has to be in a *pure* quantum state, which means a quantum state that hasn't been disturbed or affected in any way by interactions with the environment. This isn't so much of a problem with teleportation experiments that involve just photons, because photons don't interact much with the surrounding air. But similar experiments with atoms, molecules, and bigger systems have to be carried out in a vacuum to avoid collisions with gas molecules. Also, these larger objects have to be well insulated thermally from their surroundings.

The first and crucial step to a successful teleportation is entanglement. But entangled quantum states are notoriously fragile and the more atoms involved, the more fragile the states become. Not surprisingly, then, a major research goal is simply to achieve the entanglement of systems as an end in itself. Future forays into teleportation will then be able to build on this foundational work.

Early studies of entangled states tended to be restricted to elementary particles such as protons, electrons, and photons. But in 1997, a French team led by Serge Haroche at l'École Normale Supérieur in Paris gave the first demonstration of entanglement between whole atoms.[26] The experiment produced two physically separated, entangled atoms of rubidium in a superposition of states. One of these was the lowest energy or ground state and the other a much higher energy or excited state known as a circular Rydberg state, in which the outer electrons move far from the nucleus in what we can picture (in terms of classical physics) to be circular orbits. When the researchers measured one of the atoms to determine its state, they found its entangled partner always to be in the complementary state, as entanglement and its weird effect of "spooky action at a distance" demands. Their results opened the door to teleportation at the atomic level and beyond.

In 2001, Tomas Opatrny (then at F. Schiller University in Germany, now at Palacký University in the Czech Republic) and Gershon Kurzicki (of the Weizmann Institute in Israel) came up with a proposal to do just that–teleport a complete atom, including its energy of motion and all of its other external states. Their method runs like this: Using a pulse from a laser, break apart a very cold molecule into two atoms, *A* and *B*. Manipulate these two atoms so that they become entangled and allow one of the entangled atoms, say *A*, to collide with atom *C*, whose unknown state is to be teleported. After the impact, measure the momentum values of the collision partners *A* and *C*. Using the information gathered, nudge and deflect atom *B* so that its motion precisely emulates that of particle *C*. Following a complete teleportation like this, atom *B* would effectively have become *C*, and so would behave in exactly the same way. For example, if the particle-to-be-teleported *C* yielded a particular diffraction pattern after passing through two slits, then that same pattern would be produced by particle B, which received the teleported data. To carry out their procedure, said Opatrny and Kurzicki, would be "hard but feasible."[37]

An important step toward teleporting much larger objects also came in 2001 when Eugene Polzik and his colleagues at the University of Aarhus Quantum Optics Center in Denmark effectively used a beam of infrared light to tell a one-trillion-atom cloud of cesium gas to take on a particular quantum property–the quantum spin–of another cesium cloud.[29] Previously, the best that anyone had managed was the four-ion entanglement, mentioned in chapter 8, achieved by David Wineland's group at the National Institute of Standards and Technology (NIST) in Colorado.

The Aarhus demonstration was significant both because of its scale and the length of time over which it was sustained. In other experiments, individual atoms had been entangled only when they were very near together, separated by mere microns, either as ions closely spaced inside a tiny trap, or as neutral atoms flying over a short range through narrowly spaced cavities. In the Danish experiment, a pair of three-centimeter-long glass capsules into which cesium gas had been injected were positioned several *millimeters* apart. In fact, the separation distance could have been much greater but was limited so that the phials could be conveniently placed within the same magnetic shield–a crucial part of the apparatus that isolated the cesium samples from outside electromagnetic influences.

With the glass capsules in place, the researchers treated each sample with a pulse of laser light to impart a different collective spin to each atomic cloud. The collective spin describes, in effect, the net direction in which all of the atoms' tiny magnets add up. Then the team sent a single laser beam, of a carefully chosen wavelength, through both the samples. With the act of measuring the polarization of this pulse as it emerged, the spins of the two clouds were forced to entangle. A similar laser shot half a millisecond later–a geological era by quantum-mechanical standards–showed that while the orientation of each cloud's spin had shifted somewhat, the original relationship between the two clouds' orientations remained the same.

Polzik and his team didn't actually achieve full entanglement, in which the state of each particle depends on the state of every other particle. This would have lasted only a fleeting million-billionth of a second or so. Instead, they generated two weakly entangled clouds of cesium, one with slightly more atoms spinning in one direction and the other with slightly more spinning in the opposite sense. This loose interdependence of the clouds was better able to survive measurements or interactions that alter the quantum states of just a few of the constituent atoms.

Just as significant as the physical size and duration of the experiment was the ease with which the quantum information passed from the optical form (the light beam) to the physical (the gas cloud). Being able to reliably transmit the information had been a stumbling block in previous entanglement exercises. The Aarhus result was important because safe transfers will be crucial not only for far-out technologies like remote matter teleportation, but also for quantum communications. The team's use of a laser to bring about the entanglement of the disconnected clouds of atoms holds the promise for longer-distance quantum communication, which requires a set of entangled particles at each end of the quantum connection. Like kids at an egg-toss contest, the Aarhus group plans to continue to widen the gap between samples to see how far they can separate the clouds and still trigger entanglement.

Research into ways of entangling large systems of atoms and molecules can be seen as an essential first step along the road to teleporting macroscopic systems. But, in fact, the main reasons for doing this work at the moment are connected with fundamental physics or, more often, with learning techniques that may prove useful in building quantum computers. Some work relevant to quantum computing has already been done on large-scale entanglement in solids.

In 2003, for example, physicists from the Universities of Chicago and Wisconsin and University College in London, saw the effects of entanglement in the bulk properties of a magnetic material for the first time.[23] Sayantani Ghosh at the University of Chicago and his colleagues analyzed experiments that had been carried out more than a decade earlier on a crystal of a magnetic salt containing lithium, holmium, yttrium, and fluorine. Then they compared the experimental results with theoretical predictions–and made a remarkable discovery.

The holmium atoms in this salt all behave like tiny magnets and respond to each others' magnetic fields by adjusting their relative orientation, just as compass needles swing around to align themselves with the magnetic field of Earth. But in the presence of an external magnetic field, the atoms change this settled orientation. The extent to which they line up with the field is called the salt's magnetic susceptibility. Ghosh and his team looked at how the salt's susceptibility changed with temperature. Theory told them it ought to drop as the temperature climbed, because the extra energy at higher temperatures causes the atoms to jiggle around more, which upsets their ability to stay well aligned. Sure enough, the experimental data backed up this idea. But something strange happened at very low temperatures, less than a degree or so above absolute zero (the lowest temperature of all). The holmium atoms were in much better alignment than they should have been if they had normal quantum energy levels. The researchers looked at all the possible explanations for this effect and realized that only one would work: the holmium atoms had become entangled.

They uncovered a similar discrepancy between theory and experiment, at low temperatures, in the salt's ability to absorb heat–its so-called heat capacity, defined as the amount of heat needed to change the temperature of one kilogram of the substance by one degree centigrade. The way the salt's temperature rose as heat was added showed once again that entanglement was at work.

This came as a major shock. Until then, physicists had thought that naturally occurring entanglement was confined to subatomic particles. But here it evidently was, acting between atoms and remaining measurable on a macroscopic scale. The researchers say their work shows that entanglement can occur in a disordered solid that's far from perfect. They see these dense, solid-state magnets as promising systems for both fundamental quantum mechanics and potential quantum computing applications.

One physicist who wasn't surprised by what Ghosh and his group found was Vlatko Vedral, a theoretician at Imperial College in London. He'd actually predicted back in 2001 that something like this would turn up. In fact, Vedral is convinced that more research will show that entanglement is causing all sorts of significant effects in materials, and that it may even be observable someday in a substance at room temperature.

Benni Reznik, a theoretical physicist at Tel Aviv University in Israel, wouldn't need much convincing of such a possibility. According to his calculations, all of empty space–what's normally called a vacuum–is filled with pairs of particles that are entangled. Thomas Durt of Vrije University in Brussels also believes that entanglement is ubiquitous. Starting from the same basic equations that led Schrödinger to discover entanglement (well before he coined the name in the wake of the Einstein-Podolsky-Rosen paper), Durt made the astonishing discovery that almost all quantum interactions result in entanglement, whatever the conditions. "When you see light coming from a faraway star, the photon is almost certainly entangled with the atoms of the star and the atoms encountered along the way," he says. The continuous interactions between electrons in the atoms of all objects and substances ensure that everything is a spaghettilike mass of entanglements. And we humans are no exception.

Although it may be some time before we can exploit the entanglements that are apparently all around us and inside us, the prospects are astonishing. Teleportation depends on entanglement. If nature has already been kind enough to entangle large portions of itself, then it seems that a lot of the work needed to teleport macroscopic objects has already been done.

In 2002, another milestone was reached along the road to teleporting atoms, molecules, and larger bodies made from them. But this time it came purely in the form of a theory by Indian physicists Sougato Bose of the University of Oxford and Dipankar Home of the Bose Institute in Calcutta. Bose and Home showed how it might be possible to entangle *any* kind of particle. To date, physicists have only been able to entangle photons, electrons, and atoms, using different, specific methods in each case. Atoms are entangled typically by forcing them to interact inside an optical trap, photons are made to interact with a nonlinear optical crystal, and so forth. Bose and Home, however, described a single mechanism that could be used to entangle any kind of particle, even atoms or large molecules.

Think about the spin of an electron. To entangle the spins of two electrons, you first need to make sure that the electrons are identical in every respect apart from their spin. Then you fire the electrons simultaneously at a beam splitter. This forces each electron into a state of superposition, which gives it an equal probability of traveling down either of two paths. Only when you try to detect the electron do you know which path it took. If you split two electrons simultaneously, both paths could have one electron each (which will happen half the time) or either path could have both. What Bose and Home showed mathematically, is that whenever one electron is detected in each path, these particles will be entangled, even if they came from completely independent sources. The technique, they believe, should work for any object as long as the beam can be split into a quantum superposition. Anton Zeilinger has already shown that this ghostly amalgam is possible with buckyballs–molecules of fullerene–containing sixty carbon atoms. Although entangling such large objects is beyond our technical abilities at the moment, this is the first technique that might one day make it possible.

In another development in 1998, three quantum chemists at the University of California at Berkeley showed how "intraspecies" teleportation might be achieved.[35] To date, all teleportation experiments have involved teleporting quantum states between objects of the same type–photon to photon, atom to atom, and so forth. Christopher Maierle, Daniel Lidar, and Robert Harris, however, proposed a method by which information contained in the handedness, or chirality, of a molecule could be teleported to a photon. Many organic molecules come in two different forms–right-handed and left-handed–that are mirror images of one other. The Berkeley trio showed how the superpositions of these states could be transferred to a teleportee particle of light.

• • •

June 17, 2004, was a red-letter day in the history of teleportation. Two independent teams reported in the same issue of the journal *Nature* the teleportation of atoms "at the push of button."[2,30,41] It was a pivotal breakthrough on two accounts: First, it involved atoms, which are relatively huge affairs by today's teleportation standards. And second, it was *deterministic*–doable predictably upon demand.

Up until this time, as we've seen, the vast majority of teleportation experiments had centered on photons. Nonphoton experiments, like that of Haroche and his colleagues in France and also the one at Aarhus involving trillion-atom puffs of cesium, had focused simply on achieving some level of entanglement. Even the full-blown teleportation projects that were based on photons had had a probabilistic twist to them. That's to say, they relied on an element of hindsight: the supposedly teleported photon had to be detected and then have its properties checked against those of its stay-at-home entangled partner to make sure that a genuine teleportation had taken place. Unfortunately, this process of checking, inevitably involving a wave-function-collapsing measurement, wrecked the quantum state of the teleported particle. The teleportation had been verified, but now there was nothing to show for it. By contrast, the twin results on teleported atoms announced in June 2004 were deterministic in the sense that the outcome was both knowable in advance and deliberate. You effectively pressed a button, which triggered the creation of an entangled pair of atoms and the subsequent teleportation of one of them, and you could be confident ahead of time that around 75 percent of the entangled photons that arrived at Bob, the recipient, were the real teleported McCoy. This score of 75 percent is what's called the fidelity value. It's a figure of merit that gives the overall rate at which Bob receives exactly the teleported state that he is supposed to. Most previous experiments had achieved a value much lower than this.

One of the groups involved in the landmark atomic teleportation advance was David Wineland's team at NIST. Wineland and his people used an extension of the cold ion trap method that we saw in action in the previous chapter as part of their ongoing work to develop a prototype quantum computer. Their process starts by creating a superposition of quantum states in a single trapped ion of beryllium ($^9Be^+$)–an atom of beryllium-9 that is positively charged because it's missing one electron. Using laser beams, they teleported qubits describing the quantum state of the ion to a second ion with the help of a third, auxiliary ion. The properties that were teleported included the spin state of the ion (up, down, or a superposition of the two) and the phase (which has to do with the relative positions of the peaks and troughs of an ion's wave properties). Lasers were used to manipulate the ion's spin and motion, and to entangle the ions by linking their internal spin states to their external motion.

A key technical advance reported by Wineland's team was the ability to entangle the ions and then move them apart in the trap, maintaining entanglement all the while, without generating much heat. Waste heat production had previously been a problem leading to uncontrolled motions that got in the way of the process. For this latest work, the NIST group used smaller electrodes to produce electric fields that moved the ions between different zones of the multizone cold trap. Each of the NIST group's teleportations was carried out under computer control and took about four milliseconds.

The other experiment that achieved a similar result was carried out at the birthplace of teleportation, the University of Innsbruck, by a team of Austrian and American researchers led by Rainer Blatt. Like the experiment at NIST, this involved ion trapping. It relied on the same basic protocol–the auxiliary particle method that we've become familiar with–and it achieved a similar fidelity value. In the Innsbruck case, however, the ions involved were of the element calcium ($^{40}Ca^{+}$). Also, rather than moving the ions around as in the NIST technique, the Innsbruck approach was to effectively hide them temporarily in a different internal state.

Both experiments marked an important step along the way to practical quantum computing. Systems using atoms are arguably the leading candidates for storing and processing data in quantum computers, offering the most promise in terms of scalability. Also, these new experiments incorporated most of the features needed for large-scale information handling using ion traps and had the added benefit of being fairly simple in design, which would lend them to being used as part of a series of logical operations needed for computing on real-life problems.

In the 1986 remake of *The Fly*, Seth Brundle (played by Jeff Goldblum) says, "Human teleportation, molecular decimation, breakdown, reformation, is inherently purging. It makes a man a king." Unfortunately, what it makes of Brundle in the end is a disgusting insect-human hybrid. The physics of quantum teleportation, not to mention the basic principles of biochemistry, ensure that such a bizarre mixing of identities could never occur in practice. But will human teleportation, or teleportation of any living thing, ever leave the realm of science fiction?

Not surprisingly, leading scientists in the field at present play down the possibilities of applying their technique to anything much above molecular dimensions. Scientists are by nature a cautious bunch. Charles Bennett, for example, has said, "I think it's quite clear that anything approximating teleportation of complex living beings, even bacteria, is so far away technologically that it's not really worth thinking about it."

Yet, interestingly, by general consensus the problem *is* technological rather than theoretical. No known laws of physics stand in the way of quantum teleportation being applied to people. It's true that based on the technology we have today, human teleportation, or anything like it, is far from being on the drawing board. But even experts have a poor record when it comes to forecasting future trends. Sometimes we have to fall back on imagination.

Shortly after Charles Bennett and the other members of the Montreal Six laid down the first blueprint for quantum teleportation in 1993, University of Washington physicist John Cramer conjured up a scenario in which interstellar travelers take advantage of the technique in a vastly evolved form. Cramer imagines a spacecraft that sets out for the neighboring star Tau Ceti, 11.9 light-years away, carrying the macroscopic entangled states of similar states held in storage on Earth. Upon arrival after perhaps centuries or millennia of travel, robots unload the ship and set up a teleport receiver unit for each entangled state. When all is ready, the colonists step into the transmitter units where they join the stay-at-home entangled states and are destructively measured by the transmitter. The results of the measurements are recorded and sent by radio or light beam to the receivers at Tau Ceti. Twelve years later, Earth time, the beamed information is received at Tau Ceti, the transformations on the entangled states are performed, and the colonists emerge from the receiver units. For the teleported colonists, no subjective time at all has passed, and they have had a perfectly safe trip because it was known that the receiver apparatus had arrived and was fully checked out before the transmission began. The explorers can also return to Earth, with the same twelve-year gap in subjective time, using other entangled subsystems that were brought along for the trip to teleport in the opposite direction. As Cramer points out, this form of weird quantum jumping might not have quite the visceral appeal of warp-factor-nine starships, but it might be a much safer and more efficient way of exploring the stars.

No one knows where we'll be in terms of our science and technology, or what we'll be doing, fifty or a hundred years from now. Who could have imagined in 1903, the year of the Wright brothers' historic first flight at Kitty Hawk, that well within a lifetime we'd be walking on the moon, sipping champagne over the Atlantic at Mach 2, or watching as our robot probes flew past the outer planets on their way to interstellar space? Who, at the dawn of the electronic computer age in the 1940s, could have dreamed they might live to see notebook computers or the Internet? History teaches us that if something is possible and there's a strong enough motivation to make it happen, it *will* happen–and, generally, sooner rather than later.

This isn't to downplay the difficulties involved in teleporting complex, macroscopic objects. To make human teleportation a reality, scientists will need to exploit as-yet-undiscovered means of exactly duplicating the quantum states of all the particles in a human body. They'll need extraordinary devices, probably based on nanotechnology, that can deconstruct and reconstruct a system of trillions and trillions of particles, in perfect order, at astonishing speed.

"In principle, you can recreate anything anywhere, just as long as you send information on the object luminally, you have the raw materials, and you're willing to destroy the original," said Hans Christian von Baeyer, a physicist at the College of William and Mary in Virginia. "But it would take an unbelievable amount of data processing. Even a coffee cup, without the coffee, would take many times the age of the universe."

In other words, as a preliminary to complex teleportation you would need a computer vastly more powerful than any currently available on Earth. Will such a device become available? Almost certainly: such a machine is the quantum computer. The very means needed to help bring about teleportation of the large and elaborate will likely be made possible over the coming decades through mastery of teleportation of the very small and simple–the qubits on which quantum computers will run.

The advent of the quantum computer will rapidly change the outlook on what is technologically possible. First, it will enable simulations to be performed of physical systems at the quantum level. Thus it will serve as both a faithful microscope on the quantum world and a design tool for researchers striving to build the technology that will enable the entanglement and teleportation of sophisticated collections of particles. Second, large quantum computers will be able to support the "unbelievable amount of data processing," to which von Baeyer referred, as a preliminary to making unrestricted teleporting feasible.

The quantum computer is the joker in the deck, the factor that changes the rules of what is and isn't possible. Over the next decade or so, we may see the routine teleportation of individual atoms and molecules. When we have quantum computers and networks of quantum computers at our disposal, it may not be too long before we see the teleportation of macromolecules and microbes. After that, who will be prepared to bet that we ourselves won't be far behind?

10

Far-fetched and Far-reaching

Human teleportation isn't going to happen tomorrow, or next year, or, barring some miraculous breakthrough, in the next twenty years. It may never happen. But this needn't stop us from thinking about the consequences if it ever *does* become possible. If nothing else, such speculation can shed interesting light on what it means to be a human being. In fact, a number of philosophers have already used teleportation, and teleportation incidents, to delve into the mysteries of personal identity.

• • •

Imagine it's your first time to be teleported. You're standing on the pad waiting for the process to start. In a few seconds you'll be quickly dismantled atom by atom and then, an instant later, rematerialized a continent away. Everyone who's done this before tells you there's nothing to worry about. A slight tingling sensation is all you'll feel; then it's as if someone turned off the lights and a split second later turned them back on again. Some people get a bit disoriented on their first few jumps because of the sudden change of surroundings, but that's about it.

All the same, you feel uneasy. Teleportation does, after all, *destroy* the original. What appears at the other end may be a perfect copy, with not a particle out of place, but it's still a copy. It won't share a single atom in common with your old body. What bothers you is that, despite what people say, the person that's teleported won't really be you.

Look at it practically, say your friends who've made the jump. An atom is an atom; they don't have personalities or differences. Every oxygen atom is exactly like every other oxygen atom; one carbon atom's indistinguishable from any other carbon atom. (True, there are different isotopes–three of them in the case of oxygen, ^{16}O, ^{17}O, and ^{18}O–but teleportation makes sure each of these is copied correctly, too.) When you get down to the atomic and subatomic levels, nature just stamps things out identically. So it doesn't matter if all your atoms are swapped around for a different set. No test, in theory or in practice, could distinguish the teleported you from the original you.

You also have to bear in mind, as those who've gone before point out, that the stuff in your body changes anyway. Even without teleportation, atoms and molecules are continually streaming into and out of your body, so that over time every bit of matter of which you're made is exchanged. Materially, you're only in part the same person you were, say, six months ago. We're all made of matter that's from a common cosmic pool and is endlessly recycled. According to one estimate (and you have to allow a certain scientific license with these things), over the past half-billion years every atom of carbon that's now in your body has been inside another living organism an average of at least five million times. By another guesstimate, you're completely replaced at the cellular level every seven years. So teleportation, its supporters tell you, doesn't do anything qualitatively different from what happens in the normal course of events. Teleportation just happens to replace all your particles in one fell swoop rather than over a period of time.

But there's more to a person than a mere heap of atoms and molecules. Something happens when you put those tiny building blocks together in a certain way. Emergent properties appear: thoughts, memories, consciousness, personality—life itself. Yes, but those same emergent properties will arise from an identical pattern of atoms and molecules. Because quantum teleportation produces a perfect copy, right down to the subatomic level, it also, inevitably, leads to the same higher-level properties. It re-creates your brain, down to the last synapse and synaptic impulse, so that the person who steps off the pad at the other end is thinking exactly the same thought that you were, and has exactly the same set of memories that you had, when you were disassembled just before the leap. Original or copy: surely it doesn't matter what word we use as long as you feel the same and look the same, and there's no test on Earth or in the universe that can prove that you're not *in fact* the same.

Still, you're not completely happy about this business. For

one thing, whatever anyone says, teleportation involves a complete break. Normal changes to a person take place gradually–you lose and gain atoms a few at a time–so there's always a link with your previous state, a material connection with the past. When you teleport, that material link is broken. You're the same person in a figurative sense, but a completely new person in a literal sense. This discontinuity leads to two different ways of looking at human teleportation, and they're disturbingly contradictory. It's either the ultimate form of transport or a very effective way of killing someone.

Another troublesome thought is that there may be aspects of you–vital, indispensable aspects–to which the technology of teleportation is blind. The material components of our bodies and brains may well be transferred without a hitch. But there's no guarantee about any *immaterial* parts. Physics has nothing to say about souls or spirits, or the possibility of a life force or chi. These are no more than folk beliefs, unnecessary hypotheses, as far as science is concerned. Nevertheless they may be real. If they are, they may be our most essential components, and it isn't at all certain how they relate to our particular atomic patterns. What if the soul or life force is connected to matter in such a way that it remains attached if changes to the body are slow and gradual, as happen every day, but becomes separated if there's a wholesale and lightning-quick replacement as in teleportation? There aren't any soul meters to check that your spirit self made it safely across the quantum void. And asking others who've already made the trip wouldn't help, because how would they know if their souls had been stripped away or not? It might not feel any different during this life not to have a soul.

Different religions, and different schools of thought within religions, will have their own responses to such issues. In traditional Judeo-Christian theology, humans are considered to be psychosomatic entities. In other words, the body and that which animates it, referred to in Hebrew as *nephesh*, are intimately related. Each of us, according to this view, is a unique embodied soul. We come as a package: deprived of either body or soul, or separated body from soul, we're not complete. What this means in terms of human teleportation, however, isn't entirely clear.

One point of contention, even among orthodox Christians and Jews, is whether the soul can exist, even temporarily, in a state of detachment from the body like a deus ex machina. There's a long-standing tradition that to some extent it can in the form of souls in purgatory (which C. S. Lewis described as a kind of Lenten fast after death, in preparation for the great Easter of the Resurrection), the spirits of Catholic saints, various manifestations of ghosts, and so forth–although these states are regarded as only halfway houses on the path to eventual reunification. There's also a line of belief that sees the soul as an emergent property of the brain, and therefore similar to consciousness in this respect. The soul emerges at conception and supposedly grows with the growth of the body. Belief in this viewpoint can make the idea that the soul is infused or given at some point, or taken away at some point, seem to reduce the body to mere meat. And it can lead to a strong aversion to any body-soul dualism. Someone holding this antidualist position might argue that a deleted and faxed human is a different human–a different embodied soul. On the other hand, quantum teleportation is such a peculiar process that we have to be careful about jumping to any firm conclusions. If a person's body is re-created to perfection, why shouldn't the accompanying *nephesh* or soul be perfectly copied, too?

From a scientific perspective, these may seem to be silly, unfounded concerns. But if human teleportation ever does become fact, people are bound to voice fears about what it would mean for them personally. The stakes are too high not to look very closely before making this particular kind of leap. A debate at least as fierce as that over human cloning will rage on the ethical and theological implications. Perhaps there'll be a split in the Church and in the wider community, with liberal members opting to teleport while the more conservative-minded denounce the technology, arguing that anyone who uses it will literally lose their soul. There may be an age of consent: no teleportation until the age of eighteen or twenty-one, to give individuals the chance to make up their own minds about whether it's something they want to do.

In the end, the deciding factor about human teleportation for most people will likely be its sheer convenience. At first, the cost might be very high because of the enormous amount of data that has to be transferred and the complexity of the equipment needed. Early users are likely to be wealthy and adventurous–future Richard Bransons traveling by tomorrow's equivalent of the Concorde. But as the technology becomes cheaper and people gain confidence in its safety, its use will probably become as routine as catching a bus or making a phone call. With the planet's hydrocarbon reserves low or exhausted, teleportation systems (perhaps running on solar power) will offer a clean, extraordinarily handy way of crossing distances ranging from a few hundred meters to many thousands of kilometers and, eventually, beyond our planet.

Providing there's a teleportation facility at both source and destination, and relay stations along the way, there aren't any limits in principle to where people and things might be sent by this means. With a teleportation receiver on Mars, for example, perhaps set up by robotic craft, humans could simply beam themselves to that world along with the supplies and materials needed to build a base. Provided the system is kept in good working order, they could be back home on Earth in the time it takes to make a phone call. Future interplanetary explorers might even become daily commuters: an eight-to-four job mining the asteroid belt and back in time for dinner. All journeys would be made at the speed of light–a few minutes to Mars, a few hours to the outer solar system–although to the teleported person it would seem like no time at all.

Numerous issues will have to be dealt with, particularly on the security front, even if the technical challenges of human teleportation are overcome. The closest analogy we have today is the Internet and the problems that users face on a daily basis. Any global teleportation system will need the equivalent of passwords, secure sites, firewalls, and so forth to prevent people or things from beaming in when and where they're not wanted. Without safeguards, thieves could materialize inside homes in the dead of night, controlled substances and arms could be shipped around the world without detection, and terrorists could teleport bombs into government offices. Entire invasion armies could suddenly appear behind enemy lines. In fact, if teleportation of complex objects of any kind begins to loom on the technological horizon, it's quite possible that advanced developments will take place under military wraps. Xenophobic governments might ban the technology to prevent foreigners from freely entering their territories. Worldwide teleportation would quickly make a mockery of customs regulations, border checks, and immigration controls. Its introduction could bring about the final dissolution of national boundaries, so meaningless would these become. But it would also raise the specter of teleportation hackers, viruses, and spyware, and the havoc these might wreak in a world where virtually all of humanity is continually moving back and forth through the informational domain. Identity theft would take on a whole new meaning.

Only *quantum* teleportation has so far been carried out in the real world and that's the only kind of teleportation that, according to known science, can create *perfect* replicas from the subatomic level up (though approximate cloning may be possible[14, 24]). But if we're prepared to relax the rules on how exact the teleported copy has to be, we can envisage a form of teleportation that's more akin to medical scanning. Since this would work outside the pure quantum domain it would be immune from the no-cloning theorem, so that the original wouldn't need to be destroyed. Call it *classical* teleportation. It wouldn't be of any use for quantum computing or similar applications that depend critically on quantum effects such as entanglement. But it might be a viable and attractive alternative to quantum teleportation for transporting everyday objects and possibly even people. It all depends how nearly identical the original and copy have to be.

Classical teleportation could work in a similar way to the transporters in *Star Trek*. A powerful three-dimensional scanner would analyze the subject on the transporter pad and create a data matrix–a digital blueprint–that's temporarily held in a vast computer memory. Almost certainly the power of a quantum computer would be needed to handle the storage and transmission. The subject would then be dismantled, turned into some kind of stream of matter or radiation, beamed to the destination, and put back together on arrival.

Classical teleportation might even prove to be more doable than quantum teleportation when it comes to complex macroscopic objects. After all, while we still have no clue how the quantum version would work on anything large and elaborate, medical scanning technology, based on MRI (magnetic resonance imaging), CT (computerized tomography), and ultrasound, already exists and is getting more powerful and sophisticated all the time. The same is true of computers and their memories. Rematerialization might be accomplished by a kind of reverse scanner that, instead of analyzing the atomic makeup, synthesizes it using the blueprint information. This rematerialization process would presumably draw on some aspect of nanotechnology–the technique, still in its earliest stages of development today, of building things at the atomic and molecular level.

The *Star Trek* transporter sends both information *and* matter/energy. But classical teleportation, like quantum teleportation, wouldn't have to involve transmitting the subject's material contents at all. In fact, this would seem to be incredibly wasteful and impractical. To send the actual particles–molecules, atoms, electrons, and so on–of the subject would limit the speed of transmission to a fraction of the speed of light. Nothing material can reach light-speed, and even traveling at a good portion of it incurs the effect of mass increase predicted by Einstein's special theory of relativity. For example, something with a mass of 50 kilograms at rest would have a relativistic mass of nearly 115 kilograms at 90 percent of the speed of light. On top of this, there's the problem of keeping track of the matter en route and making sure none of it escapes so that there's enough original matter to enable reconstruction at the other end. Perhaps a technique could be figured out, along the lines of data compression, so that only part of the matter would have to be sent, and the rest could be filled in at the terminus. But it would still be a relatively slow and cumbersome process. On the other hand, converting the mass of an object to radiation before transmission would be even more unfeasible. The mass of a typical adult human, turned entirely into energy, would release the equivalent of about 10,000 Hiroshima bombs. This vast outpouring of radiation would somehow have to be contained, channeled to wherever it was supposed to be going, and then reconstituted into matter at the destination.

Conversion of matter to energy can be done; in fact, it's already been done in the laboratory. At the Stanford Linear Accelerator Laboratory (SLAC) in California in 1997, scientists created two tiny specks of matter–an electron and its antimatter twin, a positron–by smashing together two ultrapowerful beams of radiation. But to achieve even this small success took all the resources of a two-mile-long particle collider. It doesn't seem reasonable to consider rematerializing an ordinary large object this way, let alone a complex, fragile human being.

If classical teleportation ever becomes a reality, the likelihood is that it will involve sending information only. The difference between this and quantum teleportation, however, is that the original doesn't have to be lost in creating the copy. Information-only classical teleportation can be nondestructive–and this leads to some fascinating possibilities.

For one thing, a nondestructive teleporter is also a replicator. It scans the original in great detail and then makes a copy some distance away. Given such a capability, nothing would need to be manufactured in the ordinary way any more. Factory assembly lines would be replaced by teleporter-replicators that simply materialized any number of copies of an original design held in a memory bank. Small steps in this direction have already been made. An object can have its shape and dimensions digitized, and a robot assembler is then guided to produce a copy from an unfinished article. The digitized plans may also be sent to a remote site and be used for assembly there. A highly advanced, computer-aided design and manufacture (CADAM) system would have a lot in common with a primitive nondestructive teleporter.

At some point, the distinction between building a more or less exact copy of something to a plan and genuine teleportation becomes blurred. For example, I might read in a book how to make an origami model of a frog from a square piece of paper. I call you on the phone and tell you to get such a piece of paper and fold it this way and that until you have your own frog. Has the frog been teleported? If you say no, of course not, then what if you have an origami machine that folds the paper automatically when fed instructions. I transmit the instructions to your machine and it makes a copy of the frog that's virtually indistinguishable from my own. If that doesn't count as teleportation, what if your machine can even make its own paper, perhaps from raw elements. What if I have a machine that scans my frog to get the layout of its molecules and sends the details to your machine, which builds its own virtually identical frog, molecule by molecule. Is that teleportation?

This is more than a mere word game. The point is that nondestructive teleportation is, in a sense, already with us. We've always been able to make remote copies of things. One person creates an original, tells someone else how to do it, and he or she makes his or her own copy. Maybe this isn't teleportation as we normally think of it, but it has some of the features of teleportation. And as those features are gradually refined by technical innovations, we may come to accept that at some stage what we have is just about as good as a teleporter.

Several existing or in-development technologies seem promising as future components of a device that will effectively do nondestructive teleportation. These budding technologies include high-precision scanners, powerful (probably quantum) computers, ultra high-speed data links, and nanoscale assembly machines. An important issue is the accuracy with which the copy needs to be made. We know that classical teleportation can't make a perfect copy down to the quantum level, but that wouldn't matter in many cases. We're unlikely to be concerned if a copied widget has one of its molecules three nanometers out of place from the original, or if a single atom has gone astray from an otherwise immaculate teleported replica of a rare violin.

If nondestructive teleportation ever attains the caliber where, for example, it could make a good-as-makes-no-difference copy of a Stradivarius appear on the other side of the world, it would revolutionize commerce, end poverty, and provide more or less anything anybody wanted, anytime. Perhaps. A lot would depend on how routine the technology became, how accessible, and how portable. But the raw elements of which most things are made–oxygen, carbon, silicon, iron, aluminum, and so forth–are cheap and plentiful. If they can be assembled on demand into any desired object according to plans held in a computer memory, then through this combination of teleporter and replicator it's hard to see how we would want for anything again. Food, clothing, machines–all the other essentials and luxuries of life–would be created, copied, or teleported as desired from materials readily available in the earth, sea, and air.

A good classical teleporter, with its ability to replicate without destroying the original, would be vastly more useful for most everyday purposes than a quantum teleporter. Once scanned, an object could be replicated any number of times and its pattern held in storage until required. Break a glass? No problem, replicate another. Fancy a cordon bleu meal? Choose from a menu of thousands in your teleporter-replicator's database and have it instantly synthesized. If you like the wine that goes with it, teleport a bottle to your father across the ocean for his birthday. This is the world of *Star Trek* and of Aladdin's lamp, where teleportation and replication technologies live side by side and are used seamlessly for almost any task you can imagine.

But teleporting *people* by nonquantum means is an entirely different ball game. Even the most high-fidelity classical teleportation system isn't going to make a perfect copy of you. At the subatomic level, at least, there'll be differences. And the differences may extend higher up, to the molecular level and beyond. The issues we raised earlier in connection with quantum teleporting humans return with a vengeance. Now it isn't just a question of whether being an exact copy is as good as being the original, but how good a copy you have to be to still be you.

Obviously, there can be some differences between you and the copy. If a single hair of your head ends up out of place, it isn't going to throw your personal identity into doubt, nor is having a few molecules in your liver marginally displaced going to endanger your sense of self. But just as obviously, a teleporter that gets, say, only three-quarters of the atoms it transmits within a centimeter of where they're supposed to go will leave your transported copy not only dead but a shapeless mass of goo on the floor. The problem lies in determining what level of precision is adequate in between such extremes. A teleporter might do an otherwise very good job but consistently make a small error: for example, in positioning a particular atom in a particular nucleic acid, which could have disastrous effects later on.

But let's assume that a classical teleporter has been built that seems to work well with other living things–rats, dogs, and so forth. They arrive safely and apparently healthy. This would still leave major questions about whether it would be safe for people to use. Just delivering someone alive and well isn't good enough: that "someone" has to be you. And in this regard, what it comes down to is how accurately your brain and its present electrochemical state have to be copied for you still to be in residence upon rematerialization.

To a significant extent, we are who we are because of our memories. The synaptic pathways of the brain must be replicated well enough that these memories are left intact, otherwise the owner is lost, too. But beyond that, the brain is an organ that functions by means of billions of electrical impulses and chemical reactions every second. The precise momentary state of these, too, it would seem, must be carried over if the teleported person is to remain who he or she was when he or she set out. Ideally, the individual would be thinking the same thought upon arrival (if nothing else, to avoid disorientation). If mentally reciting a poem, for instance, at the moment of teleportation, he or she should be able to continue at the point he or she left off when the body is reassembled. We simply don't know exactly what level of detail is needed to preserve such continuity of memory, thought, and personality, but it surely must be very high.

It may even be that the level must extend to quantum depths. Neuroscientists are generally of the opinion that consciousness is an emergent phenomenon of the brain occurring at classical levels. But set against this is a minority view, expressed by Roger Penrose and others, that awareness stems from quantum processes occurring within cells. Depending on your viewpoint, you may or may not think that quantum teleportation, as opposed to classical, is essential in order for the person being transferred to remain the same.

Classical teleportation as applied to humans becomes especially interesting and controversial if it's nondestructive–in other words, if the original is retained. In this case, there would be effectively two people who could, and undoubtedly would, claim to be you. Yet how can this be? It seems impossible that both the copy and the original could be you, because you're only one person. It also seems to make sense that the "real" you is the original. Suppose, for instance, the teleported copy of you materializes on the edge of a Martian cliff and immediately plummets to his death. You might feel shocked by this, but it wouldn't harm the untransported you back on Earth in any way. On the other hand, if you were told that your teleported duplicate was fine but that you–the original who remained behind–had suffered cellular damage during the teleportation scanning process and that this would kill you in five minutes, it would be a personal tragedy. The knowledge that a more or less exact duplicate of you would remain alive and healthy on Mars wouldn't do anything to lessen the horror you would feel of your own impending doom.

These examples, in which it seems clear that you are the original and not the copy, may lead you to think that the copy couldn't be you under *any* circumstances, and that even during a destructive teleportation (classical or quantum) in which your original body is destroyed, teleporting would amount to committing suicide.

You might say I've prejudiced the argument by talking about a copy and an original. It may seem obvious that in the situation just described you are the original–the one who doesn't get teleported. But the person who materializes on Mars has no doubt that he or she is you as well! He'd feel like you and he'd rightly point out that no test on Earth (or Mars) could tell you apart (assuming the process was extremely accurate). What does this mean? Although a person can have only one body, one brain, and one selfawareness, there's nothing to stop two people having more or less identical bodies, brains, and inner thoughts.

The problem is that we're not used to thinking of cases where identity and consciousness are duplicated. It just doesn't happen in everyday life. The closest we come to it is the case of identical twins. Human cloning, if it happens, will also have some similarities. But identical twins and clones aren't perfect copies; in particular, the detailed wirings of their brains may be very different, so that they clearly have different identities. Both objectively and subjectively they're different people. But either quantum teleportation or hi-fi classical teleportation could produce essentially identical neural networks so that there really is a duplication of the brain-mind.

We can sharpen this problem in a number of ways. For example, suppose that you're nervous about teleportation and ask to be given an anesthetic before going on your first jump. While you're unconscious, the teleportation is carried out and results in an essentially identical copy of you at some remote place B while the original you remains at place A. When you come around and open your eyes, what do you see: (1) place A, (2) place B, (3) place A *and* B, or (4) neither?

To answer this, we need to broaden our ideas of personal identity and consciousness to allow for the possibility that, in principle, there really can be more than one person who has a right to call him or herself you. Only option 4 of those listed isn't correct. The creation of a duplicate you can't deprive both the people at A and B of conscious experience. Option 3 is true in the sense that there are now two individuals who look and think like you. But options 1 and 2 are also correct, because although identity and consciousness can be duplicated, they can only be experienced in the singular. The person at A wouldn't also see through the eyes of the identical person at B, and vice versa.

You might object to this analysis and say that the person who remained behind at A is the "real" you. The person who was teleported to B may call him or herself you, but he or she is in fact a copy. If it hadn't been for the anesthetic, there'd have been no break in A's conscious, unteleported experience and there'd be no doubt who was the original. But then what if we teleport two copies (which we can do since this is a classical device), to places B and C and destroy the so-called original at A. Which of the people at B and C is now the real you?

One of the difficulties any future human teleportation may engender is this kind of prejudice about who is a copy and who is an original. "Placism" could be rampant unless steps are taken to stamp it out from the very start. A large part of the problem comes from the potential of classical teleportation to make copies nondestructively. It may be necessary to outlaw nondestructive teleportation in the specific case of human beings; otherwise, not only would Earth's population mushroom out of control, but all sorts of ethical and legal issues would arise. My copy teleports while I remain home. Then my copy returns and demands that I share my home, spouse, and children with him. Worse, while he was gone, he generated additional copies of himself, and now they're all wanting a room in my house and separate checkbooks for my bank account. And what if one of my replicas–another me–gets a parking ticket? Do I have to pay? Do we split the cost? How could I prove it wasn't me when in another sense it really *was* me?

To avoid such unpleasantness, classical human teleportation will surely have to be used strictly Star Trek–style: the teleportee shimmering out in one place and reappearing somewhere else. This "one person out, one person in" rule would prevent Earth from becoming clogged with numerous replicas, and we would avoid endless debates and disputes about who was really who.

But in the *Star Trek* universe, the transporter sometimes malfunctions or other circumstances conspire and two copies are made. In an episode of *Star Trek: The Next Generation* called "Second Chances," an identical copy of commander Wil Riker is created. We're even given a bit of technobabble about how it happens. Years ago, while a then-lieutenant Riker was beaming up from a planet's surface through severe atmospheric interference, the transporter chief locked on to Riker's signal with a second tracking beam. When this second tracking beam turned out not to be needed, it was abandoned–but not lost. Unbeknownst to everyone on the ship, the ionic disturbance in the atmosphere caused the second beam to be reflected back to the planet and result in the creation of a second Riker. Fast forward eight years and the two Rikers meet. Confusion reigns, Riker-2 gets together with Riker-1's old girlfriend before matters are resolved, and Riker-2 departs to pursue his separate existence.

Perhaps, despite precautions, such accidental duplication will occasionally happen in the future world. A play that aired on BBC radio in 2004 raised a more disturbing possible consequence of this. A young woman is scheduled to go to Australia for a business meeting and checks in at the teleportation travel center. There's a technical glitch during the transfer and she's told that she'll have to come back and try again a different day. Feeling a bit shaken from the experience, she goes home. A few days later, she's congratulated at work on her successful meeting in Australia and keeps running into other people who speak as if she'd done things she hadn't. Of course, it turns out that during the transporter glitch she was duplicated. Matters now become more sinister because it transpires that the teleportation company has a team to take care of such situations by doing manually what the teleporter is supposed to do automatically: destroy the original person. This aspect of the operation is normally kept secret as it's considered not very good for business. The unfortunate woman spends the rest of the story trying to avoid the hired assassins.

Would such disposal of the original, which future law might insist be a routine part of the teleportation process, be considered murder? By today's standards, undoubtedly yes. But ethics may change in a world intent on avoiding multiple copies of personal identity and streams of consciousness. Common law at present regards murder as the act of killing *plus* malice aforethought. This might change if there comes a time when it's considered unacceptable to have more than one you in existence. Killing the original may then be considered simply a way of disposing of an unnecessary and troublesome copy–a surplus use of flesh and bone. After all, the teleported you would no doubt claim that, despite the loss of life, you're actually still alive and well, with all memories intact. And he or she might be relieved to hear that there'd be no rival claims on possessions, partner, and so forth.

Although it might not be desirable to have more than one you running around at any one time, there's no reason why a backup copy of you couldn't be retained in the teleporter-replicator's memory. Then if anything happened to the living you, the backup could be used as a replacement. This would be a shortcut to immortality or to time travel into the future. Tired of the present world? Just teleport yourself into oblivion for a century or so and be replicated at some preprogrammed future date. Or, as a form of health insurance, have all your organs digitized while you're still young and healthy, so that as your parts wear out you can have the previously stored replicas materialized and transplanted–or even surgically teleported directly into your body.

One way or another, teleportation is going to play a major role in all our futures. It will be a fundamental process at the heart of quantum computers, which will themselves radically change the world. Some form of classical teleportation and replication for inanimate objects also seems inevitable, and this, too, will have dramatic effects. Whether humans can ever make the impossible leap remains to be seen. One thing is certain: if that impossible leap turns out to be merely difficult–a question of simply overcoming technical challenges–it will someday be accomplished.

Epilogue

Luk closed the book and leaned back. Strange that electronic books had never caught on. Sometimes old technologies were simply the best and couldn't be improved upon. Other times . . . He looked out across the blue waters of Port Jackson to the gleaming towers and spires of the great southern city and wondered how many of its inhabitants had already made that strange, seemingly impossible leap this morning to almost anywhere on the planet.

A gay array of cruise boats and windsurfers was crisscrossing the Harbor, an endless movement to and fro. What was he, where was he, when he was en route somewhere between trans-pads? And what was he when he rematerialized? A consciousness re-formed thanks to a coming together of matter–a coming together exactly like his "old" self that a moment earlier had been taken apart and unceremoniously dumped into a bunch of containers? When you thought about what even "solid" matter actually was made of, it seemed incredible that thought, life, and structure of any kind was possible.

Three rising tones. The signal that some*thing* was arriving on one of the trans-pads in the apartment; a person materializing triggered a different sound. It would be the groceries he'd ordered the day before, a couple of boxes of produce conveniently beamed in from the local store. Except, of course, it wasn't local at all. With teleportation, distance was literally no object. These vegetables, canned goods, and other provisions may have come from Thailand, France, or anywhere else in the wink of an eye. You placed your order on the Internet with the supplier of your choice and they delivered–not to your door, but to your very kitchen–or whatever trans-pad coordinates you gave. The stuff came, quality guaranteed, from any of untold numbers of giant warehouses dotted around the globe, each equipped with thousands of trans-pads. It was the same with clothing, appliances, and furniture. Whatever you wanted was ordered and paid for online and dispatched through thin air into your home, a few minutes or a few hours later.

It was getting harder to remember what the world had been like before teleportation. Having to cart things around. Having to travel everywhere physically. Until a few years ago, there'd still been plenty of planes flying because of the teleportation age limit. No one who was pregnant or under eighteen was allowed to use the trans-pads because of the "soul issue" (except in a medical emergency with parental consent). It was argued that people should have the chance to make a reasoned ethical choice before they made their first jump. But that law had been repealed. The whole process had become so routine: billions of jumps every day and no one ever seemed the worse for having his or her molecules taken apart. Convenience had won the day. So now there were no restrictions. Families could teleport together, without having a babysitter beam in to watch the kids. Children could teleport to school and go on exotic field trips.

The airlines, except for one or two that catered to the tiny minority of "originals," had gone out of business. Oddly enough, ocean cruises had become all the rage. People, it seemed, still occasionally needed to travel in old-fashioned style and at a leisurely pace. And it was true, teleporting had taken some of the romance and adventure out of life. You could go anywhere in the world that had a trans-pad at the drop of a hat: Machu Picchu at sunset, Mount Everest base camp, you name it. The trouble was, it was just as quick and easy for everyone else. So you had to share the romantic sunset with ten thousand others who'd had the same bright idea as you.

Luk's fifth anniversary with his wife was coming up and Luk had been racking his brain for somewhere really exotic to take Rosie. The Bahamas? A night in Monte Carlo? And then it came to him. The moon! That new hotel on the rim of Copernicus crater. It was a pricey jump, to be sure. You got routed via one of the teleport satellites in Earth orbit, to a similar relay in lunar orbit, and then down to the surface. Bandwidth was limited, the tariffs high. But this was a special occasion. He'd spring the news on her over cocktails that evening.

Later, both dressed for a night on the town (Beirut, as it happened), Luk and Rosie stood side by side ready to make the not-so-impossible leap. There was the familiar low hum that came as the trans-pad began its disassembly sequence. And, just at that moment, a fly landed on Luk's cheek. Too late to swat it away now. A distant memory stirred in his mind–an image from an old movie he'd once seen.

Uh-oh, thought Luk, as his form shimmered out of sight.

Chronology

1900	Energy found to be quantized; birth of old quantum theory (Planck).
1905	Light found to be quantized; photoelectric effect explained (Einstein).
1923	Wave-particle duality extended to matter (de Broglie).
1925	Matrix formulation of quantum mechanics (Heisenberg, Born, Jordan).
1926	Schrödinger equation and wave mechanics (Schrödinger).
	Copenhagen interpretation (Bohr, Heisenberg).
1927	Uncertainty principle (Heisenberg).
	Start of Bohr-Einstein debates on foundations of quantum mechanics.
1931	The term *teleportation* coined (Charles Fort in his book *Lo!*).
1935	EPR paradox (Einstein, Podolsky, and Rosen).
	The term *entanglement* first used (Schrödinger).
1948	Information theory (Shannon).
1952	Hidden variables theory (Bohm).
1957	Many-worlds interpretation (Everett).

1964	Bell's theorem and inequality.
1967	*Star Trek* introduces fictional teleportation to millions of viewers.
1970	First ideas on quantum cryptography (Wiesner). Not published until 1983.
	Decoherence theory.
1978	Foundations of quantum computing (Deutsch).
1979	Bennett and Brassard combine Wiesner's ideas with public-key cryptography.
1980	Quantum computing can be done coherently (Benioff).
1982	No-cloning theorem (Wooters and Zurek).
	Experiments disprove local hidden variables (Aspect et al.).
1984	First protocol for quantum key distribution, BB84 (Bennett and Brassard).
1985	Universal computer operating on quantum principles could simulate any physical process (Deutsch).
1989	Experimental prototype of BB84 protocol (Bennett, Brassard, et al.).
1991	Theory of quantum cryptology based on entanglement (Ekert).
1992	"Qubits" named (Schumacher).
1993	Theory of quantum teleportation (Bennett, Brassard, Crépau, Jozsa, Peres, Wooters).
1994	Quantum algorithm for efficient factorization of large numbers (Shor).
1996	2-qubit quantum computer.
1997	First practical demonstrations of teleporation

(Zeilinger et al., Innsbruck and DeMartini et al., Rome).

1998 Teleportation of complete particle (Kimble et al., Caltech).

NMR teleportation (Laflamme).

1999 3-qubit quantum computer.

2000 NMR used to create 7-qubit quantum computer (Los Alamos).

2001 Quantum entanglement of cloud of cesium atoms (Polzik et al.).

2002 Quantum key sent across 23 km of open air (UK Defence Research Agency).

Entanglement shown possible for any kind of particle (Bose and Home).

2004 Decoherence observed in the lab (Zeilinger et al.).

Teleportation of quantum states of atoms.

First commercial hardware to support quantum cryptography.

First electronic transfer of funds using entangled photons.

First quantum cryptography network (Qnet).

Teleportation distance record increased to 500 meters.

2015 ? Teleportation of molecules.

2020 ? First quantum computer.

2050 ? Teleportation of a virus

2100+ ? Human teleportation

References

1. Aspect, A., P. Grangier, and G. Roger. Experimental tests of realistic local theories via Bell's theorem *Physical Review Letters* 47 (1981): 460.
2. Barrett, M. D., J. Chiaverini, T. Schaetz, J. Britton, W. M. Itano, J. D. Jost, E. Knill, C. Langer, D. Leibfried, R. Ozeri, and D. J. Wineland. Deterministic quantum teleportation of atomic qubits. *Nature* 429 (2004): 737–39.
3. Beenakker, C. W. J., C. Emary, M. Kindermann, and J. L. van Velsen. Proposal for production and detection of entangled electron-hole pairs in a degenerate electron gas. *Physical Review Letters* 91 (2003): 147901.
4. Beenakker, C. W. J., and M. Kindermann. Quantum teleportation by particle-hole annihilation in the fermi sea. *Physical Review Letters* 92 (2004): 056801.
5. Bell, J. S. On the Einstein-Podolsky-Rosen paradox. *Physics* 1 (1964): 195–200.
6. Bennett, C. H, F. Bessette, G. Brassard, L. Salvail, and J. Smolin. Experimental quantum cryptography. *Journal of Cryptology* 5 (1992): 3–28.
7. Bennett, C. H. and G. Brassard. Quantum cryptography: Public key distribution and coin tossing. *Proceedings of the IEEE International Conference on Computers, Systems, and Signal Processing*. New York: IEEE Press (1984): 175.
8. Bennett, C. H., G. Brassard, C. Crépeau, R. Jozsa, A. Peres, and W. H. Wootters. Teleporting an unknown

quantum state via dual classical and Einstein-Podolsky-Rosen channels. *Physical Review Letters* 70 (1993): 1895–1899.

9. Bennett, C. H., G. Brassard, and N. D. Mermin. Quantum cryptography without Bell's theorem. *Physical Review Letters* 68 (1992): 557.
10. Blinov, B. B., D. L. Moehring, L.-M. Duan, and C. Monroe. Observation of entanglement between a single trapped atom and a single photon. *Nature* 428 (2004): 153–57.
11. Bohm, D. A suggested reinterpretation of quantum theory in terms of hidden variables. *Physical Review* 85 (1952): 611–23.
12. Boschi, D., S. Brance, F. DeMartini, L. Hardy, and S. Popescu. Experimental realisation of teleporting an unknown pure quantum state via dual classical and Einstein-Podolsky-Rosen channels. *Physical Review Letters* 80 (1998): 1121–25.
13. Bouwmeester, J. W., J. W. Pan, K. Mattle, M. Eibl, H. Weinfurter, and A. Zeilinger. Experimental quantum teleportation. *Nature* 390 (1997): 575–79.
14. Buzek, V., and M. Hillery. Quantum cloning. *Physics World* 14 (11): 25–29 (2001).
15. Cirac, J. I. Quantum physics: Entangled atomic samples. *Nature* 413 (2001): 375–77.
16. Deutsch, D. Quantum theory, the Church-Turing principle and the universal quantum computer. *Proceedings of the Royal Society of London, A* 400 (1985): 97–117.
17. Diffie, W. and M. E. Hellman. New Directions in Cryptography. *IEEE Transactions on Information Theory* 22 (1977): 644–54.
18. Duan, L. M., M. D. Lukin, J. I. Cirac, and P. Zoller. Two experiments towards the implementation of a quantum repeater. *Nature* 414 (2001): 413.

19. Dürr, S., T. Nonn, and G. Rempe. Origin of QM complementarity probed by a "which-way" experiment in an atom interferometer. *Nature* 395 (1998): 33.
20. Einstein, A., B. Podolsky, and N. Rosen. Can a quantum-mechanical description of physical reality be considered complete? *Physical Review* 47 (1935): 777–80.
21. Ekert, A. K. Quantum crytography based on Bell's theorem. *Physical Review Letters* 67 (1991): 661–63.
22. Furusawa, A., J. L. Sorensen, S. L. Braunstein, C. A. Fuchs, H. J. Kimble, and E. S. Polzik. Unconditional quantum teleportation. *Science* 282 (1998): 706–9.
23. Ghosh, S., T. F. Rosenbaum, G. Aeppli, and S. N. Coppersmith. Entangled quantum state of magnetic dipoles. *Nature* 425 (2003): 48–51.
24. Grosshans, F., and P. Grangier. Quantum cloning and teleportation criteria for continuous quantum variables. *Physical Review A* 64 (2001): 010301/1–4.
25. Hackermüller, L., K. Hornberger, B. Brezger, A. Zeilinger, and M. Arndt. Decoherence of matter waves by thermal emission of radiation. *Nature* 427 (2004): 711–14.
26. Hagley, E., X. Maitre, G. Nogues, C. Wunderlich, M. Brune, J. M. Raimond, and S. Haroche. General of Einstein-Podolsky-Rosen pairs of atoms. *Physical Review Letters* 79 (1997): 1.
27. Hartley, R. Transmission of Information. *Bell System Technical Journal* 7 (1928): 535–63.
28. Itano, W. M., D. J. Heinzen, J. J. Bollinger, and D. J. Wineland. Quantum Zeno effect. *Physical Review A* 41 (1990): 2295–2300.
29. Julsgaard, B., A. Kozhekin, and E. S. Polzik. Experimental long-lived entanglement of two macroscopic quantum objects. *Nature* 413 (2001): 400–403.
30. Kimble, H. J., and S. J. Van Enk. Quantum physics: Push-button teleportation. *Nature* 429 (2004): 712–13.

31. Kok, P., C. P. Williams, and J. P. Dowling. Construction of a quantum repeater with linear optics. *Physical Review A* 68 (2003): 022301.
32. Kuzmich, A., W. P. Bowen, A. D. Boozer, A. Boca, C. W. Chou, L.-M. Duan, and H. J. Kimble. Generation of non-classical photon pairs for scalable quantum communication with atomic ensembles. *Nature* 423 (2003): 726.
33. Laflamme, R., M. A. Nielsen, and E. Knill. Complete quantum teleportation using nuclear magnetic resonance. *Nature* 396 (1998): 52–55.
34. Marcikic, I., H. de Riedmatten, W. Tittel, H. Zbinden, and N. Gisin. Long-distance teleportation of qubits at telecommunication wavelengths. *Nature* 421 (2003): 509–13.
35. Mariele, C. S., D. A. Lidar, and R. A. Harris. How to teleport superpositions of chiral amplitudes. *Physical Review Letters* 81 (1998): 5928–31.
36. Nyquist, H. Certain topics in telegraph transmission theory. *Bell System Technical Journal* 7 (1928): 617–44.
37. Opatrny, T., and G. Kurizki. Matter-wave entanglement and teleportation by molecular dissociation and collisions. *Physical Review Letters* 86 (2001): 3180–83.
38. Pan, J.-W., S. Gasparoni, M. Aspelmeyer, T. Jennewein, and A. Zeilinger. Experimental realisation of freely propagating teleported qubits. *Nature* 421 (2003): 721–25.
39. Peres, A. and W. H. Wootters. Optimal detection of quantum information. *Physical Review Letters* 66 (1991): 1119–22.
40. Popescu, S. Teleportation versus Bell's inequalities. What is nonlocality? *Physical Review Letters* 72 (1994): 797.
41. Riebe, M., H. Häffner, C. F. Roos, W. Hänsel, J. Benhelm, G. P. T. Lancaster, T. W. Körber, C. Becher, F. Schmidt-Kaler, D. F. V. James, and R. Blatt. Deterministic quantum teleportation with atoms. *Nature* 429 (2004): 734–37.
42. Schrödinger, E. Die gegenwärtige situation in der

quantenmachanik (The present situation in quantum mechanics). *Die Naturewiffenschaften* 23 (1935): 807–12, 824–28, 844–49; English translation in: *Proceedings of the American Philosophical Society* 124 (1980): 323–38.

43. Schrödinger, E. Discussion of probability relations between separated systems. *Proceedings of the Cambridge Philosophical Society* 31 (1935): 555–63.
44. Scully, M. O., B. G. Englert, and H. Walther. Quantum optical tests of complementarity. *Nature* 351 (1991): 111–16.
45. Shannon, C. E. A Mathematical theory of communication. *Bell System Technical Journal* 27 (Jul. and Oct. 1948): 379–423 and 623–56. Reprinted in N. J. A. Sloane and A. D. Wyner, eds. *Claude Elwood Shannon: Collected Papers.* New York: IEEE Press (1993).
46. Shannon, Claude E. Communication theory of secrecy systems. *Bell System Technical Journal* 28 (4) (1949): 656–715.
47. Taylor, G. I. Interference fringes with feeble light. *Proceedings of the Cambridge Philosophical Society* 15 (1909): 114–15.
48. Smolin, J. A. The early days of experimental quantum cryptography. *IBM Journal of Research and Development.* 48 (1): 47–52.
49. Wiesner, S. Conjugate coding. *Sigact News* 15 (1): 78–88 (1983). Original manuscript 1970.
50. Wootters, W. K., and W. H. Zurek. A single quantum cannot be cloned. *Nature* 299 (1982): 802–03.
51. Young, Thomas. *A Course of Lectures on Natural Philosophy and the Mechanical Arts.* London: J. Johnson (1807).
52. Zeilinger, A. A foundational principle for quantum mechanics. *Foundations of Physics* 29 (4) (1999): 631–43.

Bibliography

Bell, J. S. *Speakable and Unspeakable in Quantum Mechanics.* New York: Cambridge University Press, 1993.

Bohr, Niels. *Essays: 1958–1962 on atomic physics and human knowledge.* New York: Interscience, 1963.

Bouwmmester, D. et al., eds. *The Physics of Quantum Information: Quantum Cryptography, Quantum Teleportation, Quantum Computation.* Berlin: Springer-Verlag, 2000.

Brown, Julian. *Quest for the Quantum Computer.* New York: Touchstone Books, 2001.

Deutsch, D. *The Fabric of Reality.* New York: Penguin, 1997.

Greenberger, D., Reiter, L., and Zelinger, A., eds. *Epistemological and Experimental Perspectives on Quantum Mechanics.* Boston: Kluwer Academic Publishing, 1999.

Gribbin, John. *Schrödinger's Kittens and the Search for Reality.* Boston: Bay Back Books, 1996.

Hirvensalo, Mika. *Quantum Computing.* 2nd ed. New York: Springer-Verlag, 2004.

Johnson, George. *A Shortcut through Time: The Path to the Quantum Computer.* New York: Knopf, 2003.

Milburn, Gerard J. *Schrödinger's Machines: The Quantum Technology Reshaping Everyday Life.* New York: W. H. Freeman, 1997.

Nielsen, Michael A. and Chuang, Isaac L. *Quantum Computation and Quantum Information.* New York: Cambridge University Press, 2000.

Pais, Abraham. *Niels Bohr's Time: In Physics, Philosophy, and Polity.* Oxford: Clarendon Press, 1991.

Parfit, Derek. *Reasons and Persons.* Oxford: Oxford University Press, 1984.

Peres, Asher. *Quantum Theory: Concepts and Methods (Fundamental Theories of Physics, Vol. 57).* Boston: Kluwer Academic Publishers, 1995.

Scully, M. O., and Zubairy, M. S. *Quantum Optics.* New York: Cambridge University Press, 1997.

Singh, Simon. *The Code Book: The Evolution of Secrecy from Mary, Queen of Scots to Quantum Cryptography.* New York: Doubleday, 1999.

Turton, Richard. *The Quantum Dot: A Journey into the Future of Microelectronics.* New York: Oxford University Press, 1996.

Wheeler, J. A. and Zurek, W. H., eds. *Quantum Theory and Measurement.* Princeton: Princeton University Press, 1983.

Index